Algebraic Equations of Linear Elasticity

Lester W. Schmerr Jr.

Algebraic Equations of Linear Elasticity

Novel Force-based Methods for Solid Mechanics with MATLAB®

Springer

Lester W. Schmerr Jr.
Iowa State University
Woodward, IA, USA

ISBN 978-3-031-66176-1 ISBN 978-3-031-66174-7 (eBook)
https://doi.org/10.1007/978-3-031-66174-7

This Springer imprint is published by the registered company Springer Nature Switzerland AG
The registered company address is: Gewerbestrasse 11, 6330 Cham, Switzerland

Preface

Solid mechanics is primarily concerned with the forces acting on and within structures. Those forces are related through the requirement that they be in "equilibrium." However, the conditions of equilibrium are often not sufficient by themselves to solve for the forces. Such problems are called statically indeterminate problems. The additional equations needed can be obtained by recognizing that the deformations caused by the forces are not independent but satisfy certain compatibility equations. If those deformations are related to the forces, then the compatibility equations become the additional relations between the forces that, when combined with the equilibrium equations, allow us to solve for the forces. In simple problems one can obtain the compatibility equations from diagrams of the changes of the geometry due to the forces. However, for more realistic structural problems it was found that the use of displacements and energy methods was more effective. These combinations of equilibrium equations with compatibility equations to solve statically indeterminate problems form the basis of what we will call the *first generation of force-based methods*.

In the 1960s the growing power of computers and their ability to solve problems with matrix algebra methods led to the development of an alternative method where the displacements of a structure were related to the known applied forces through a stiffness matrix. Compatibility equations were not required since the use of displacements guaranteed that compatibility of the deformations was satisfied. It became clear that a displacement-based approach where a structure is modeled as a collection of elements whose stiffness matrices can be defined, allowed the solution of highly complex structures. This approach was called the *displacement-based or stiffness-based finite element method.*

Unfortunately, the first generation of force-based methods could not be easily implemented with the same computer-based matrix formulations used in the displacement-based finite element method so the first-generation force-based method quickly faded, and the displacement-based finite element method became the method of choice. Today, powerful finite element software tools are available as are numerous books and tutorials on the finite element method.

This dominance of the displacement-based finite element method has had a profound effect on the teaching of courses in solid mechanics in an engineering curriculum. Since forming stiffness matrices with computer-based matrix algebra typically requires a relatively advanced engineering and programming background, first and second level engineering courses such as statics and strength of materials largely ignore finite elements, which is taught only at an advanced undergraduate or graduate level, usually in a separate course. In statics, only statically determinate problems are solved and in strength of materials bar and beam problems are solved by relating internal forces and moments to displacements but without the use of stiffness matrices. In advanced strength of materials courses, structures such as frames are usually considered with displacements and energy methods like those used in first-generation force-based methods.

An engineering curriculum in solid mechanics currently is highly centered around the use of displacements for solving statically indeterminate problems. This is unfortunate since engineers are most interested in forces and moments, which are obtained in displacement-based methods as secondary quantities, found only after the displacements have been obtained. There is, however, an alternative approach which we will describe in this book, where forces and moments can be found directly and then the displacements obtained as secondary quantities. This new approach we call a *second-generation force-based method*. Unlike the first-generation force-based methods, the second-generation method can be implemented at all levels in an engineering curriculum, from statics to a force-based finite element method that is as powerful as the current displacement-based finite element method and is just as suitable for computer-based implementation with matrices and vectors.

The use of the symbolic and numerical capabilities of MATLAB is an important part of including force-based methods in solid mechanics courses. With MATLAB, students can implement second-generation force-based methods at many levels in the curriculum, including statics, without any programming, by using MATLAB as a "calculator" containing all the necessary functions. A knowledge of matrix-vector algebra is needed to implement the second-generation force-based methods effectively, but that knowledge requirement is rather modest and of great value to engineering students since matrices and vectors play important roles in modern engineering practice.

As shown in this book, the second-generation force-based method emerges from a general formulation where the partial differential equations of linear elasticity are replaced by equivalent algebraic equations. These *algebraic equations of linear elasticity* can be used to solve statically indeterminate problems in reduced forms that define either the new second-generation force-based approach or a new displacement-based approach. Both reduced forms rely on only two fundamental building blocks: (1) the equilibrium equations and (2) the flexibilities of the deformable bodies. In the new displacement-based method the stiffness matrix is formed from equilibrium and flexibility matrices. In the new force-based approach the equilibrium equations are augmented by compatibility equations, which are found from homogeneous solutions of the equilibrium equations. A flexibility matrix is used to define those compatibility equations in terms of forces/moments.

The second-generation force-based method is the primary focus of this book, although the new displacement-based method is also described. Because it deals directly with forces and moments, the second-generation force-based method can be implemented even at the level of a statics course, where force and moments are the focus of the course. This allows one to treat, for the first time, statically indeterminate problems for deformable bodies in a statics course. A new text, L.W. Schmerr, *Engineering Statics with MATLAB*, CRC Press, 2024, includes such problems.

In strength of materials and advanced strength courses, the inclusion of the new force-based method gives an alternate path for solving the problems found in those courses and lays the foundations of many of the ideas of the finite element method. For example, the text, L.W. Schmerr, *Advanced Mechanics of Solids*, Cambridge Univ. Press, 2021, demonstrates the use of second-generation force-based finite element methods at the level of an advanced strength of materials course.

In elasticity courses, the derivation and use of the algebraic equations of linear elasticity can show how the difficulties of dealing with partial differential equations may be sidestepped by transforming those equations into algebraic equations with work-energy concepts.

Finally, in a finite element course a force-based finite element method can be described along with the traditional displacement-based approach to demonstrate how the two methods provide alternative ways for solving complex structural problems.

This book is not a textbook. It is a resource where engineering educators, advanced students, practicing engineers, and researchers can learn a new approach to solving statically indeterminate problems and pass that knowledge on to students and colleagues. This book derives the algebraic equations of linear elasticity and uses those equations to develop the second-generation force-based method and a new displacement-based method for solving statically indeterminate problems. Separate chapters for statics, strength of materials, advanced strength of materials, and finite elements describe these new methods at a technical level appropriate to each of those courses. Force-based solutions are obtained not by forming stiffness matrices but by guaranteeing equilibrium at the nodes of a structure, like the method of pins used by statics students to solve truss problems, and then combining those equilibrium equations with their homogeneous solutions and a knowledge of the flexibilities of the structure to obtain a solution. MATLAB is used effectively as a software tool in a manner that does not require a knowledge of programming. A second-generation force-based method implemented with MATLAB symbolic algebra is a combination that is simple and direct and offers a valuable new perspective on how to solve statically indeterminate problems.

Woodward, IA, USA Lester W. Schmerr Jr.

Contents

Chapter 1
Introduction

Objectives

- To describe how second-generation force-based methods can enhance a solid mechanics curriculum.
- To introduce the governing differential equations of linear elasticity and the principles of virtual work and complementary virtual work.
- To introduce the algebraic equations of linear elasticity and show their reduction to equations involving only the forces or only the displacements. The formulation in terms of forces is the second-generation force-based method.
- To present the elements of a first-generation force-based method.

1.1 Background

Engineers have long sought to analyze the equilibrium of structures to determine the forces present. With the development of the equations of linear elasticity for small deformations, the requisite equations were available in differential form for the stresses, strains, and displacements. However, the inability to solve these differential equations except in simple cases led to the development of algebraic methods that relied instead on work-energy principles. Such formulations received a major boost in the 1960s with the development of displacement-based finite element methods. In fact, the displacement-based finite element method has largely overshadowed an earlier force-based algebraic method based on work-energy principles (which we will call a *first-generation force-based method*) because of the generality of a displacement approach and its suitability for being implemented with matrices on computers. A comprehensive and fascinating history of the various approaches used to solve equilibrium problems is given in Kurrer [1].

L. W. Schmerr Jr., *Algebraic Equations of Linear Elasticity*,
https://doi.org/10.1007/978-3-031-66174-7_1

In this book, we will show how a *second-generation force-based method* can once again be at the forefront in the analysis of engineering structures and how the method can be used at multiple levels in an engineering curriculum. This force-based method will be shown to be based on a general formulation where a general set of governing *algebraic equations of linear elasticity* can be developed that directly correspond to all the differential equations and differential relations found in the theory of elasticity. These governing algebraic equations, like the traditional displacement-based finite element method, will be developed from work-energy principles but where stresses and displacements are treated on an equal footing instead of a purely displacement-based approach, as found in most discussions of finite elements.

From these general equations one can then develop reduced formulations that involve only the stresses (or equivalently the forces) or only displacements. These reduced formulations are also algebraic equivalents of their differential equation counterparts. The displacement formulation is similar but not identical to that of the conventional finite element formulation. The second-generation force-based method that emerges is notable for three reasons: (1) forces rather than displacements are often the quantities of most interest in an engineering analysis; (2) this force-based method, unlike earlier methods, is as suitable for computer implementation as are displacement-based methods; and (3) the force-based formulation can serve as the basis for teaching students how to solve equilibrium problems directly for the forces present at all levels in an engineering curriculum.

Figure 1.1 shows the major solid mechanics courses in an engineering curriculum and their current topics. Figure 1.2 shows the changes possible with the introduction of the second-generation force-based methods. In statics, for example, one can incorporate for the first time significant statically indeterminate problems, such as statically indeterminate trusses, as shown in Chap. 3. Other simple statically indeterminate problems, such as wire-supported structures and structures with a combination of rigid members and springs, can also be dealt with at the technical level of a statics course. This addition of statically indeterminate problems to statics is made possible because a second-generation force-based method only requires the equilibrium equations and the flexibility of the deformable elements (truss members, wires, springs) to determine the forces, so it is a simple extension of the topics familiar to statics students. The force-based method is best implemented with the use of matrices and vectors, so some additional background in matrix-vector algebra is needed. But these requirements are modest and the use of software tools such as MATLAB, where matrices play an essential role, makes the inclusion of matrix-vector topics in statics even easier. Also, by using the symbolic algebra tools of MATLAB, statically indeterminate problems can be solved without programming since MATLAB can be used as a modern engineering "calculator" where a student only needs to describe the equilibrium equations and the flexibilities and use the built-in functions of the MATLAB calculator to solve for the forces.

In strength of materials, statically indeterminate problems for the extension and torsion of members are normally considered using displacement-based methods and differential equations since the governing ordinary differential equations, in contrast

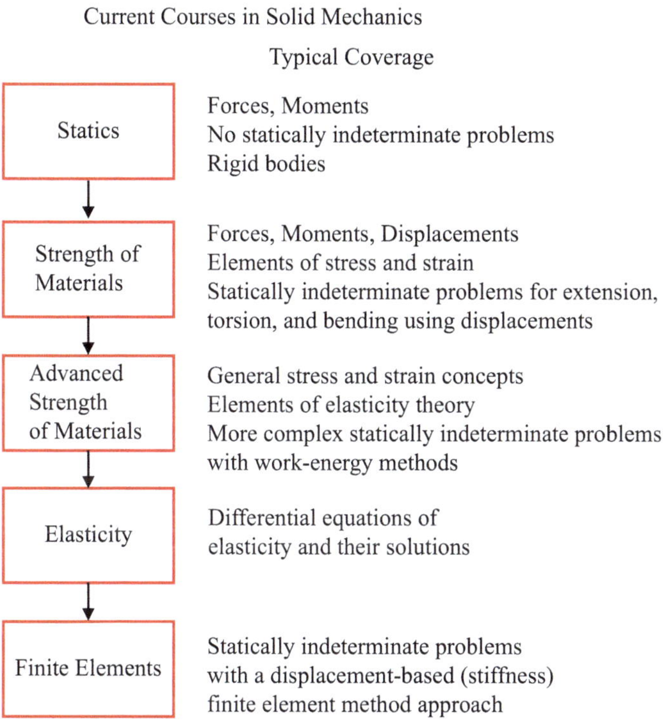

Fig. 1.1 Current topics in solid mechanics courses in an engineering curriculum

to the partial differential of elasticity, can be easily solved for the displacements. By introducing a second-generation force-based method, these problems can be solved directly for the forces/torques in a form that again only uses equilibrium and flexibility of the members, as seen in Chap. 4. The bending of beams can also be solved with a second-generation force-based method, but the detail needed to describe the method is such that bending problems are considered in this book at the level of an advanced strength of materials course in Chap. 5. Once the elements of the force-based methods are in place, however, it will be shown in Chap. 5 how bending problems can be treated with force-based methods using only topics that are familiar to strength of materials students. The use of a second-generation force-based approach in strength of materials can also introduce some basic foundations of finite element methods at the technical level of a strength of materials course without having to deal with the many details appropriate to a higher-level course in finite elements.

In advanced strength of materials, beams and more complex statically indeterminate structures such as frames under combined axial extension, torsion, and bending are normally considered with work-energy concepts and simple forms of first-generation forced-based methods. These problems can also be solved with second-generation force-based methods using only equilibrium and flexibility concepts (see

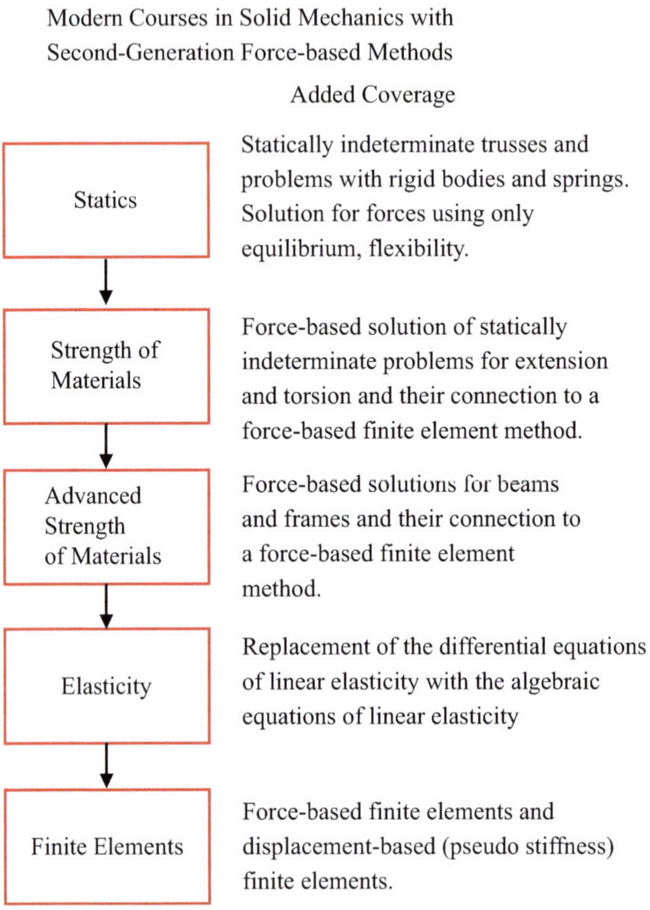

Fig. 1.2 Enhanced courses with second-generation force-based methods

Chap. 5). This builds on the strength of materials topics and again introduces the foundations of finite elements without the complicated details.

In an elasticity course, one normally treats the solutions possible with the differential equations of linear elasticity and covers the more general uses of work-energy ideas. By introducing the equivalent algebraic equations of linear elasticity into this course, as developed in Chap. 2, one can provide the foundation for solving elasticity problems algebraically with a second-generation force-based method or with a new displacement-based method.

Finite element courses traditionally have considered displacement-based approaches using stiffness matrices to solve complex problems. Forces are only calculated in this approach from the displacements in the post-processing phase of the analysis. By implementing a force-based finite element approach (see Chap. 6), one can solve directly for the forces and then determine the displacements, if needed, in the post-processing.

Current engineering solid mechanics courses are highly displacement oriented when dealing with statically indeterminate structures. This is understandable given the importance of displacement-based finite element method, which has been implemented in powerful finite element codes for analyzing the equilibrium of complex structures ranging from skyscrapers to bridges and airframes. Forces, the quantities that engineers are often most interested in, are relegated to derived quantities. Second-generation force-based methods can serve to bring the focus back on the forces and provide the basis for a new generation of modern solid mechanics courses.

1.2 Differential Equations of Linear Elasticity

The fundamental differential equations of linear elasticity are (1) the equations of equilibrium, (2) the strain-displacement relations, and (3) the stress-strain equations [2]. In matrix-vector form these equations are:

$$\begin{aligned} &[\mathcal{L}_E]\{\sigma\} + \{f\} = \{0\} \\ &\{e\} = [\mathcal{L}_E]^T\{u\} \\ &\{\sigma\} = [C]\{e\} \quad \text{or} \quad \{e\} = [D]\{\sigma\} \end{aligned} \tag{1.1}$$

where

$$\{\sigma\} = \begin{bmatrix} \sigma_{xx} & \sigma_{yy} & \sigma_{zz} & \sigma_{yz} & \sigma_{xz} & \sigma_{xy} \end{bmatrix}^T \tag{1.2a}$$

$$\{e\} = \begin{bmatrix} e_{xx} & e_{yy} & e_{zz} & \gamma_{yz} & \gamma_{xz} & \gamma_{xy} \end{bmatrix}^T \tag{1.2b}$$

$$\{u\} = \begin{bmatrix} u_x & u_y & u_z \end{bmatrix}^T \tag{1.2c}$$

are the stress components, engineering strain components, and displacement components, respectively, written as column vectors. The differential equilibrium matrix operator is

$$[\mathcal{L}_E] = \begin{bmatrix} \partial/\partial x & 0 & 0 & 0 & \partial/\partial z & \partial/\partial y \\ 0 & \partial/\partial y & 0 & \partial/\partial z & 0 & \partial/\partial x \\ 0 & 0 & \partial/\partial z & \partial/\partial y & \partial/\partial x & 0 \end{bmatrix} \tag{1.3}$$

and the body force vector is

$$\{f\} = \begin{bmatrix} f_x & f_y & f_z \end{bmatrix}^T \tag{1.4}$$

The matrix [C] is the 6 × 6 matrix of elastic constants and the matrix $[D] = [C]^{-1}$ is the corresponding compliance matrix. There are a total of 15 equations in Eq. (1.1) in terms of 15 unknowns (six stresses, six strains, and three displacements).

These governing equations are not, however, solved in this general form. If, for example, one places the stress-strain and strain-displacement relations into the equilibrium equations, one arrives at the system of equations:

$$[\mathcal{L}_E][C][\mathcal{L}_E]^T\{u\} + \{f\} = \{0\} \tag{1.5}$$

which are Navier's equations for the displacements (for a general anisotropic medium). This is a differential equation displacement formulation. Equation (1.5) represents three equilibrium equations in terms of the three unknown displacements. If the displacements are found from solutions of these equations with appropriate boundary conditions, then the stresses can be obtained from the stress-strain and strain-displacement relations.

If we wish to obtain a set of equations for the stresses only, then we need to eliminate the displacements. We expect this is possible since the six strains are obtained through derivatives of only three displacements. Thus, there must be some integrability conditions that permit the displacements to be obtained from the strains. Those integrability conditions are called the *compatibility equations*, which are given by [2]

$$[\mathcal{L}_S]\{e\} = \{0\} \tag{1.6}$$

where

$$[\mathcal{L}_S] = \begin{bmatrix} 0 & -\partial^2/\partial z^2 & -\partial^2/\partial y^2 & \partial^2/\partial y\partial z & 0 & 0 \\ -\partial^2/\partial z^2 & 0 & -\partial^2/\partial x^2 & 0 & \partial^2/\partial x\partial z & 0 \\ -\partial^2/\partial y^2 & -\partial^2/\partial x^2 & 0 & 0 & 0 & \partial^2/\partial x\partial y \\ 0 & 0 & 2\partial^2/\partial x\partial y & -\partial^2/\partial x\partial z & -\partial^2/\partial y\partial z & \partial^2/\partial z^2 \\ 2\partial^2/\partial y\partial z & 0 & 0 & \partial^2/\partial x^2 & -\partial^2/\partial x\partial y & -\partial^2/\partial x\partial z \\ 0 & 2\partial^2/\partial x\partial z & 0 & -\partial^2/\partial x\partial y & \partial^2/\partial y^2 & -\partial^2/\partial y\partial z \end{bmatrix} \tag{1.7}$$

If we write the compatibility equations in terms of the stresses through the stress-strain relations, then our governing equations become

$$\begin{aligned} [\mathcal{L}_E]\{\sigma\} + \{f\} &= \{0\} \\ [\mathcal{L}_S][D]\{\sigma\} &= \{0\} \end{aligned} \tag{1.8}$$

where $[D] = [C]^{-1}$ is the elastic compliance matrix (see Eq. (1.1)). The equations in Eq. (1.8) are in terms of the stresses only. This stress formulation, however, is nine equations for the six stresses, so we have too many equations in Eq. (1.8) for a well-posed stress formulation. To overcome this problem, one can write the stresses in terms of stress potentials which automatically satisfy the equilibrium equations, thus reducing the problem to one where the number of equations and unknowns are the same. The Airy stress function is a well-known example of this type that can be used to solve two-dimensional problems [2]. It should be pointed out that the equilibrium and compatibility matrix operators satisfy the following relations

$$\begin{aligned} [\mathcal{L}_S][\mathcal{L}_E]^T &= [0] \\ [\mathcal{L}_E][\mathcal{L}_S]^T &= [0]^T \end{aligned} \tag{1.9}$$

which, as we will see, are important parts of a second-generation force-based method.

1.3 Work-Energy Relations

If one wants to avoid having to solve partial differential equations, then we need to transform the differential equations into algebraic equations. This is possible because we can use work-energy relations to develop the principles of virtual work and principle of complementary virtual work. These principles are equivalent to satisfying the equations of equilibrium and the strain-displacement relations. The principal of virtual work, in matrix-vector form, can be written as [2]:

$$\int_V \{\delta e\}^T\{\sigma\}dV = \int_V \{\delta u\}^T\{f\}dV + \int_S \{\delta u\}^T\left\{T^{(\mathbf{n})}\right\}dS \tag{1.10}$$

where $\{\delta u\}$ are virtual changes of the displacements and $\{\delta e\}$ are corresponding virtual changes of the strains. The vector $\{T^{(\mathbf{n})}\}$ represents the stress (traction) vector acting on the surface S of a volume V whose outward unit normal is $\mathbf{n}$. In a similar fashion, we can use the strain-displacement relations and equilibrium to prove the principle of complementary virtual work given by [2]:

$$\int_V \{\delta\sigma\}^T\{e\}dV = \int_V \{\delta f\}^T\{u\}dV + \int_S \left\{\delta T^{(\mathbf{n})}\right\}^T\{u\}dS \tag{1.11}$$

where $\{\delta\sigma\}$ are virtual changes of the stresses and $\{\delta f\}$, $\{\delta T^{(\mathbf{n})}\}$ are virtual changes of the body forces and surface stress vector, respectively. Equations (1.10) and (1.11) together with the stress-strain relations

$$\{\sigma\} = [C]\{e\} \tag{1.12}$$

are the work-energy equations equivalent to the differential equations of Eq. (1.1).

1.4 Algebraic Equations of Linear Elasticity

It will be shown in Chap. 2 how the work-energy principles just presented lead to a set of equivalent algebraic equations for linear elastic problems that are replacements for the differential equations of Eq. (1.1):

$$\begin{aligned} [E]\{F\} &= \{P\} \\ \{\Delta\} &= [E]^T\{U\} \\ \{\Delta\} &= [G]\{F\} \end{aligned} \tag{1.13}$$

where the stresses are represented in term of discrete "generalized forces" $\{F\}$ (they are called generalized forces because in many structural problems they represent internal forces or moments). The matrix $[E]$ is an equilibrium matrix, and the vector $\{P\}$ is a set of generalized forces acting at a set of discrete locations called nodes. The vector $\{U\}$ is a set of discrete generalized displacements (linear displacements or rotations) at these nodes, and the vector $\{\Delta\}$ is a set of discrete deformations that replace the strains in the original governing equations so they can be thought of as generalized strains. The deformation-force equations $\{\Delta\} = [G]\{F\}$ are equivalent to the stress-strain relations, where the deformations are written in terms of forces in terms of a flexibility matrix $[G]$. We can also write the forces in terms of the deformations as $\{F\} = [G]^{-1}\{\Delta\}$, so we see that the flexibility matrix $[G]$ plays the same role as the elastic compliance matrix $[D]$ and $[G]^{-1}$ plays the same role as the matrix of elastic constants $[C]$. If there are N equilibrium equations in terms of M forces, then there are also M deformations and N displacements so that Eq. (1.13) is 2 M + N equations for 2 M + N unknowns. Like the governing differential equations, these algebraic equations can be written in reduced forms. We will consider both reduced displacement and force formulations.

1.4.1 Displacement Approach and the Pseudo Stiffness Matrix

If, for example, one writes the deformation-displacement relationship in terms of the forces as

$$\{F\} = [G]^{-1}\{\Delta\} = [G]^{-1}[E]^T\{U\} \tag{1.14}$$

and places it into the equations of equilibrium, then the equilibrium equations become

$$[E][G]^{-1}[E]^T\{U\} = \{P\} \tag{1.15}$$

Equation (1.15) is a linear algebraic equivalent of Navier's equations, Eq. (1.5). We can also express Eq. (1.15) in terms of a *pseudo stiffness matrix* $[K_p]$, where

$$[K_p] = [E][G]^{-1}[E]^T \tag{1.16}$$

and Eq. (1.15) becomes

$$[K_p]\{U\} = \{P\} \tag{1.17}$$

which, if there are N equilibrium equations in $[E]$, gives N equations for N displacements. The stiffness matrix here is called a pseudo stiffness matrix, since in Chap. 2 it is shown that it can differ from the "ordinary" stiffness matrix found in finite elements. For all the extensional, torsional, and bending problems considered in this book, however, the ordinary stiffness matrix and the pseudo stiffness matrix will be identical. Equation (1.17) can serve as the basis for obtaining the displacements. Once those displacements are found, we can obtain the forces $\{F\}$ from Eq. (1.14).

Solving problems for displacements is the usual way in which many problems are treated, particularly those that are statically indeterminate. However, we can also develop a second-generation force-based approach from the algebraic equations as follows.

1.4.2 Second-Generation Force-Based Approach

In statically determinate problems, the number of unknown forces is equal to the number of equilibrium equations present so we can solve for the unknown forces directly from equilibrium. But if, as is often the case, we have more unknown forces than equations of equilibrium, the problem is statically indeterminate, and we need to augment the equilibrium equations with additional equations. Let M equal the number of forces and let N < M be the number of equilibrium equations. In the equilibrium equations $[E]\{F\} = \{P\}$, the equilibrium matrix, therefore, will be a N × M matrix. The deformation-displacement relations then represent M equations for the deformations in terms of N displacements, so there must be M-N relations between the deformations. These M-N relations are called the compatibility equations, which we will write as

$$[S]\{\Delta\} = \{0\} \tag{1.18}$$

where $[S]$ is the (M-N) × M compatibility matrix. Written in terms of the forces, we have

$$[S][G]\{F\} = \{0\} \tag{1.19}$$

which are the additional M-N equations we need to be able to solve for the forces. The combination of the equilibrium equations for the forces and the compatibility equations in Eq. (1.18) are, therefore,

$$\begin{aligned} [E]\{F\} &= \{P\} \\ [S][G]\{F\} &= \{0\} \end{aligned} \tag{1.20}$$

which are equivalent to Eq. (1.8) except now these are M equations in M unknowns, so they are directly solvable. Thus, if we simply augment the equilibrium equations with these compatibility equations, we have:

$$\begin{bmatrix} [E] \\ - \\ [S][G] \end{bmatrix} \{F\} = \begin{Bmatrix} \{P\} \\ - \\ \{0\} \end{Bmatrix} \tag{1.21}$$

which we will write as

$$[E_{sys}]\{F\} = \{P_{sys}\} \tag{1.22}$$

which is now a solvable system of M equations for the M unknown forces. The key, of course, is to be able to determine the compatibility matrix $[S]$. From the deformation-displacement relations, we must have

$$[S]\{\Delta\} = [S][E]^T\{U\} = \{0\} \tag{1.23}$$

which can be satisfied for arbitrary displacements if

$$[S][E]^T = [0] \tag{1.24}$$

Taking the transpose, we find

$$[E][S]^T = [0]^T \tag{1.25}$$

Note that these are algebraic counterparts of the matrix differential relations of Eq. (1.9) described earlier.

Equation (1.25) says that the columns of the transpose of the compatibility matrix are force solutions $\{F_h\}$ of the homogeneous equilibrium equations, i.e.

$$[E]\{F_h\} = \{0\} \tag{1.26}$$

and there will be M-N such solutions present in the compatibility matrix $[S]$. Finding such homogeneous solutions is straightforward so we can form up and solve Eq. (1.21) to obtain the forces directly. Once the forces are found, we can also determine the displacements from those forces. The starting point is the deformation-displacement relation, which when written in terms of the forces is:

$$[G]\{F\} = [E]^T\{U\} \tag{1.27}$$

We can extend this relationship without changing it by writing it as:

$$[G]\{F\} = [E]^T\{U\} = \left[\, [E]^T \quad | \quad [G]^T[S]^T \,\right] \begin{Bmatrix} \{U\} \\ - \\ \{0\} \end{Bmatrix} \tag{1.28}$$

since the additional terms added to the $[E]^T$ matrix are multiplied by zeros so that they do not affect the original deformation-displacement relation. Equation (1.28) can also be written as

$$[E_{sys}]^T \begin{Bmatrix} \{U\} \\ - \\ \{0\} \end{Bmatrix} = [G]\{F\} \tag{1.29}$$

or, equivalently

$$[E_{sys}]^T\{U_{sys}\} = [G]\{F\} \tag{1.30}$$

where

$$\{U_{\mathrm{sys}}\} = \begin{Bmatrix} \{U\} \\ - \\ \{0\} \end{Bmatrix} \tag{1.31}$$

Once a force-based solution is obtained we already have $[G]$, $[E]$, $\{F\}$ and $[E_{\mathrm{sys}}]$ so Eq. (1.30) can be solved for the displacements $\{U\}$ and one or more zeros in $\{0\}$. The number of zeros contained in $\{0\}$ is equal to the number of compatibility equations, or equivalently, the number of rows in the compatibility matrix $[S]$. Note that the additional zeros do not represent any physical displacements present in a problem. All the real displacements (some of which may also be zero) are contained in $\{U\}$.

Equation (1.21) is a second-generation force-based method for solving statically indeterminate problems. We only need the equilibrium equations, $[E]\{F\} = \{P\}$, and the flexibility matrix, $[G]$, to implement the solution since we can determine the compatibility matrix, $[S]$, from homogeneous solutions of the equilibrium equations. Once the forces are found, Eq. (1.30) can be used to obtain the displacements.

1.5 First-Generation Force-Based Approach

We have called the force-based method described in the previous section a second-generation method for solving statically indeterminate problems because there is an earlier method that was used frequently before the rise of displacement-based finite element methods. In this section, we will briefly outline this method, using a simple example. Consider the beam shown in Fig. 1.3a which is supported by four rollers.

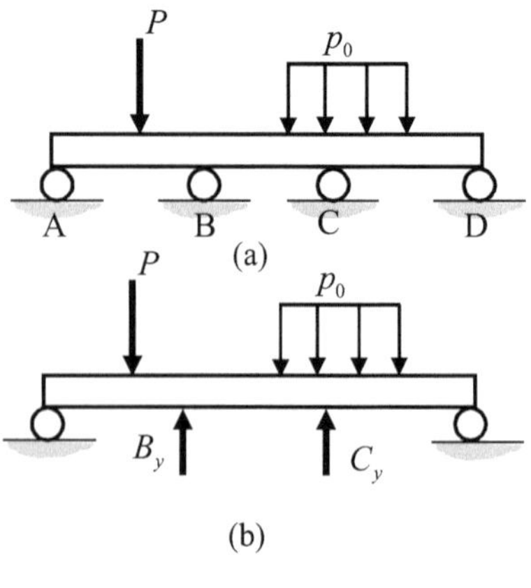

Fig. 1.3 (**a**) A statically indeterminate beam. (**b**) The redundant reaction forces at supports B and C

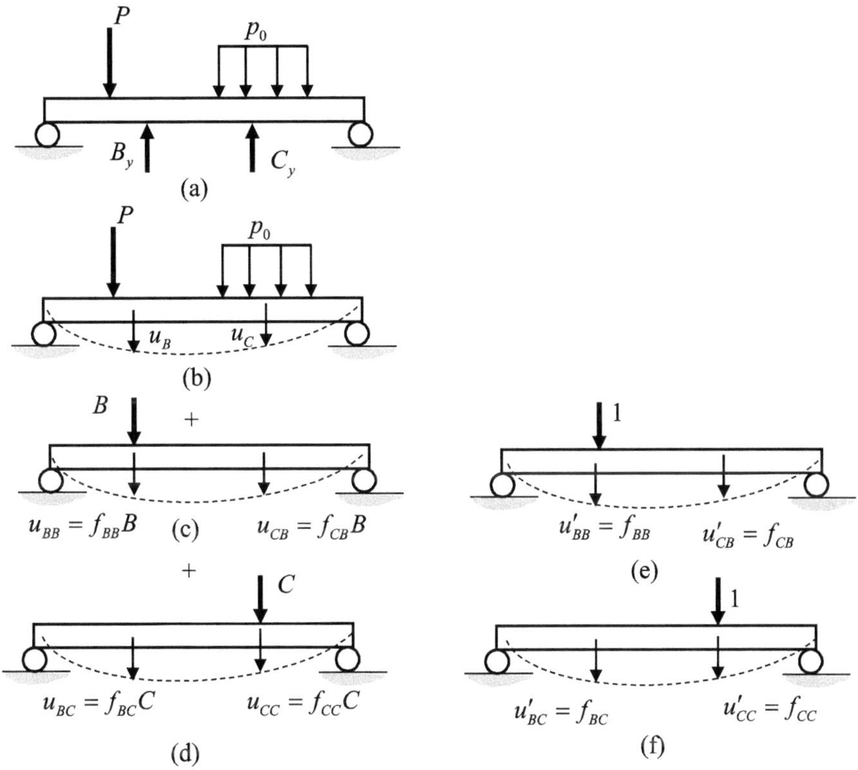

Fig. 1.4 (**a**) The applied forces and redundant reactions for the problem of Fig. 1.3 (**a**). This problem can be treated as the superposition of the problems shown in (**b**), (**c**), and (**d**), where $B = -B_y$, $C = -C_y$. The flexibilities appearing problems in (**c**) and (**d**) can be calculated from the deformations appearing in (**e**) and (**f**), respectively

Since there are only two equations of equilibrium for the beam and a total of four unknown reactions, there are two more unknown forces in this problem than equations. We could take the two "excess" forces as the reactions at B and C. These are called redundant forces (see Fig. 1.3b). Now, consider the original problem in Fig. 1.4a as the superposition of the three problems shown in Fig. 1.4b–d, where $B = -B_y$, $C = -C_y$. The forces B and C will be treated as if they were positive applied forces acting downward since rollers can only support such forces. Later, we will set them equal to their proper negative values. The problem in Fig. 1.4b is a statically determinate problem where we can compute the moment and displacement anywhere in the beam, including the displacements at points B and C, (u_B, u_C). Similarly, we can determine the moment and displacement anywhere in beam shown in Fig. 1.4c in terms of the force B. Let the displacements at points B and C in this problem be written as $u_{BB} = f_{BB}B,\ u_{CB} = f_{CB}B$, as shown in Fig. 1.4c. For the beam shown in Fig. 1.4d, we can write $u_{BC} = f_{BC}C,\ u_{CC} = f_{CC}C$. The constants f_{mn} appearing in these displacements are called the flexibilities. These flexibilities can be calculated explicitly from the unit forces acting on the beams in Figs. 1.4e, f since the beam problems in those figures are also statically determinate. Now consider adding up all the displacements at *B* and *C* and setting the sum equal to zero, since in our original problem, there were no displacements at pins *B* and *C*. We find

$$\begin{aligned} u_B + f_{BB}B + f_{BC}C &= 0 \\ u_C + f_{CB}B + f_{CC}C &= 0 \end{aligned} \tag{1.32}$$

which we can write in matrix-vector form in terms of the original forces (B_y, C_y) as:

$$\begin{bmatrix} f_{BB} & f_{BC} \\ f_{CB} & f_{CC} \end{bmatrix} \begin{Bmatrix} B_y \\ C_y \end{Bmatrix} = \begin{Bmatrix} u_B \\ u_C \end{Bmatrix} \tag{1.33}$$

Equation (1.33) can be solved for the unknown reaction forces at points *B* and *C* and so it is called a force-based method. Once the reactions are found, the original problem, therefore, becomes a statically determinate problem and the internal bending moment and displacement can be found anywhere in the beam. In this problem, we solved for the redundant reactions, but the same process can be used in structures where there are too many internal forces to be found from the equations of equilibrium alone. We will not show such cases here. The equations appearing in Eq. (1.33) are called compatibility equations since they require that the redundant forces satisfy conditions that make those redundant forces compatible with the zero displacement conditions at the supports. In the second-generation force-based method, we also need to augment the equilibrium equations with the compatibility equations $[S][G]\{F\} = 0$, which play a similar role as Eq. (1.33). The matrix appearing in Eq. (1.33) is called a flexibility matrix. In the second-generation force-based method, we defined a flexibility matrix $[G]$, which is different from the one appearing in Eq. (1.33). Since in this book we will not deal with the first-generation force-based method, this should not cause any confusion. In general, the flexibility matrix in Eq. (1.33) is symmetric, i.e., $f_{BC} = f_{CB}$, so here we must calculate only three flexibilities. This symmetry is due to the principle of reciprocity, which is satisfied for all linear elastic systems [2]. In practice, work-energy principles are often used to determine the flexibilities and calculate displacements, such as those appearing in Eq. (1.33). For a history of such force-based methods, see [1].

The first-generation force-based method is an attractive way to solve statically indeterminate problems by hand for problems that are not too large. However, it is not easily automated for computer calculations, so the displacement-based finite element method, which is highly suited to computer implementation for complex problems, has largely replaced this force-based approach. The second-generation force-based method discussed in this book can give new life to solving statically indeterminate problems directly for the forces.

1.6 About This Book

This introduction has described the algebraic equations of elasticity and their reduction to displacement-based and second-generation force-based formulations for solving statically indeterminate problems.

In Chap. 2, the algebraic equations of elasticity will be derived in detail from work-energy principles. The second-generation force-based method and a new displacement-based method that arises from the algebraic equations of linear elasticity will be used to solve several examples of statically indeterminate problems. This will give an explicit illustration of the steps needed and the issues that arise when obtaining solutions.

Chapter 3 will apply the second-generation force-based method to wire-supported structures and trusses that are like the problems normally considered in a statics course but where the problems are statically indeterminate. The symbolic algebra and numerical capabilities of MATLAB will be used to illustrate how these problems can be solved without needing programming skills, making them appropriate to the technical level of a statics course (see [3], which is a new statics text that includes second-generation force-based solutions).

Chapter 4 will apply second-generation force-based methods to problems involving the extension and torsion of members, as found in a strength of materials course. A method like the method of pins (joints) for trusses will be used where the equilibrium equations will be satisfied at the discrete nodes in the geometry. Beam problems, which are also normally considered in a strength of materials course, will be left to Chap. 5 since the added complexities of force-based solutions of beam problems are best described at a higher level. The connection of the approach used to solve axial extension and torsion problems to that of a force-based finite element method will also be discussed in Chap. 4 to introduce some of the basic ideas in a finite element approach. Force-based methods for strength of materials problems can be found in [4], where they are called the integrated force method.

Chapter 5 will apply second-generation force-based methods to beam problems and to more complex frames and structures that are in combinations of bending, axial loads, and torsion. Again, the equilibrium equations will be obtained from the forces and moments acting at discrete nodes and this approach will be connected to that of finite elements. A summary will also be given in Chap. 5 of how to include force-based solutions to beam problems in a strength of materials course which only introduces concepts covered at the level of that course, thus avoiding the complex details covered in Chap. 5. In [2], force-based methods are discussed extensively at the advanced strength of materials level, often using finite element approaches.

Chapter 6 will give a very gentle introduction to both force-based and displacement-based finite element methods and the programming steps needed to implement those methods. Only one-dimensional problems will be considered, allowing us to discuss some of the key issues in finite elements without the complicated details better left to complete finite element courses. MATLAB scripts

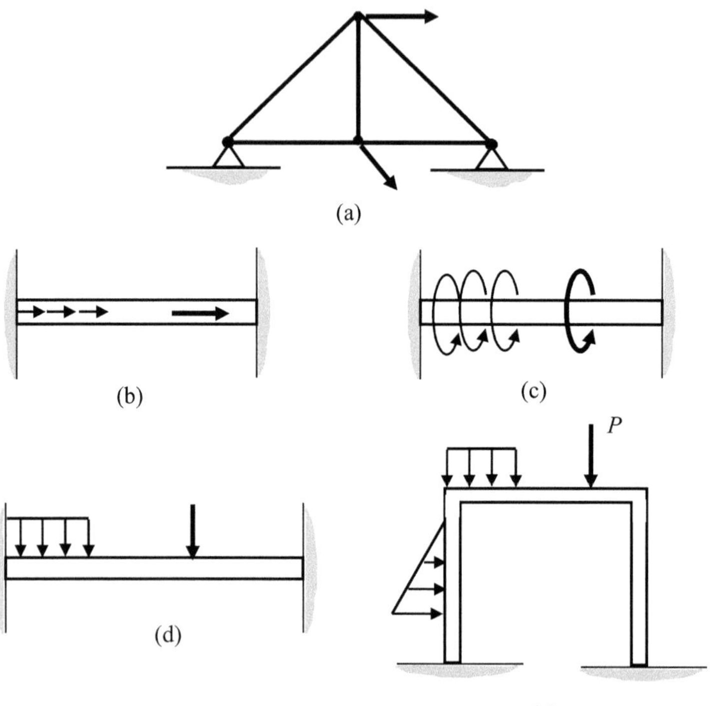

Fig. 1.5 (**a**) A statically indeterminate truss. (**b**) Statically indeterminate extension of a bar. (**c**) Statically indeterminate twisting of a bar. (**d**) Statically indeterminate bending of a beam. (**e**) Statically indeterminate extension and bending of a frame

that describe the programming steps needed will be given. See [2] where second-generation force-based finite element methods are also discussed.

By using the second-generation force-based method, all the statically indeterminate problems in solid mechanics can be treated in a unified fashion. Figure 1.5 shows examples of some typical statically indeterminate problems. The examples shown are two-dimensional problems, but three-dimensional problems can be considered in the same manner. Each of the five problems in Fig. 1.5 can be solved directly for the forces/moments by breaking the geometry into elements and replacing the applied forces/moments by equivalent forces/moments acting at the end nodes of those elements. This results in the configurations described in Fig. 1.6 where the elements are shown as dotted members, and the nodes are shown as black circles, rectangles, and squares. There are only two elements for the problems in Fig. 1.6b–d and one element for each straight-line segment for the frame of Fig. 1.6e, but these choices are not essential, and more elements could be used or, if convenient, elements of different sizes may be chosen [Note: it is necessary that at least two elements be used in the fixed-fixed extension, torsion, and bending problems given in Fig. 1.5 to ensure there is at least one node that can displace and cause the

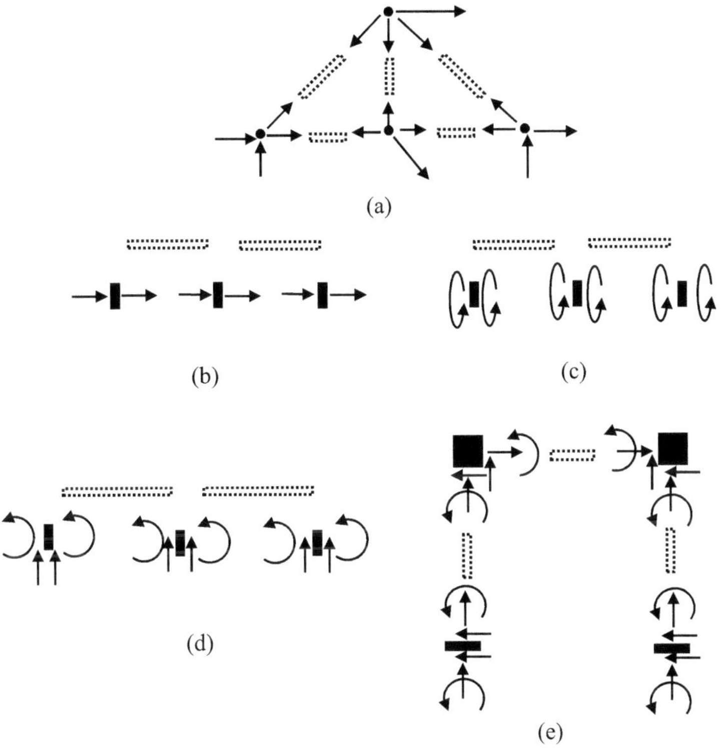

Fig. 1.6 The structures of Fig. 1.5 broken into elements where the actual applied forces/moments are replaced by equivalent forces/moments acting at the end nodes of the elements and displayed with the other forces acting at the nodes. The elements themselves are shown as dotted lines and the nodes as black circles, rectangles, or squares

adjacent elements to deform]. By applying equilibrium conditions at all the nodes and defining the flexibilities of the elements, the compatibility equations can be obtained and combined with the equilibrium equations to solve directly for all the forces/moments. This is very similar to a finite element approach except that the equilibrium equations here are written symbolically for each of the nodes rather than for each element. The built-in functions of MATLAB are then used to transform these symbolic algebraic equations to vector-matrix equations and implement the remaining steps in the solution. The elements themselves need not be considered in this approach since the internal forces in the elements all automatically satisfy homogeneous equilibrium equations. In a force-based finite element solution, on the other hand, the equilibrium equations for the entire structure are assembled from element equilibrium matrices through a specially programmed set of steps, just as done in the traditional displacement-based finite element method with element stiffness matrices. When this assembly process is implemented with computer programming it can be used to treat very complex problems without writing down the equilibrium equations at the nodes themselves, which depend on the nature of the problem and the geometry being considered. Examples of a force-based finite element approach are shown in Chap. 6. However, there is no computer programming needed when writing the equations of equilibrium for the nodes symbolically, so the problems typically treated in statics, strength of materials, and advanced strength of materials can all be considered at a technical level appropriate to those courses, and one can introduce many of the concepts of finite element methods without the added complexity and programming of a traditional finite element approach.

References

1. Kurrer, K.-E. (2018). *The history of the theory of structures: Searching for equilibrium* (2nd ed.). Ernst & Sohn.
2. Schmerr, L. W. (2021). *Engineering statics with MATLAB*. Cambridge University Press.
3. Schmerr, L. W. (2024). *Engineering statics—A matrix-vector approach with MATLAB*. CRC Press.
4. Patnaik, S. N., & Hopkins, D. A. (2004). *Strength of materials: A unified theory*. Elsevier.

Chapter 2
Algebraic Equations of Linear Elasticity and Their Solutions

Objectives

- To derive a set of algebraic equations that correspond to the governing differential equations of linear elasticity.
- To use the algebraic equations of linear elasticity to define a force-based approach where compatibility equations are obtained from solutions of homogeneous equilibrium equations.
- To use the algebraic equations of linear elasticity to define a displacement-based approach using a pseudo stiffness matrix.
- To solve for the forces and displacements in problems using force-based and displacement-based methods.

2.1 Overview

Despite the equations of linear elasticity being available for over a hundred years, solutions of these equations are difficult to obtain. Recognizing this fact, engineers have instead used a variety of algebraic methods to solve structural problems. These have typically been based on work-energy principles. In this chapter, we will show how those work-energy principles can lead to the algebraic equations of linear elasticity that were given, without proof, in the previous chapter. As seen in Chap. 1, these algebraic equations of linear elasticity can be reduced to formulations where one solves directly for either the forces or the displacements. A displacement approach based on the algebraic equations of linear elasticity is similar and, in many cases, identical to the finite element method in common use today. However, the force-based approach, called here a second-generation force-based method, is not widely known, or used. In this and subsequent chapters, we will only use this second-generation method and not deal with the first-generation force-based method

L. W. Schmerr Jr., *Algebraic Equations of Linear Elasticity*,
https://doi.org/10.1007/978-3-031-66174-7_2

described briefly in Chap. 1, so henceforth we will often refer to the second-generation force-based method as simply the force-based method. An introduction to the implementation of this force-based method using MATLAB will be given in this chapter. Later chapters will solve a wide variety of problems with a very similar set of steps. A distinguishing feature of this MATLAB approach will be the use of symbolic and numerical algebra which will allow students to use MATLAB as a modern engineering "calculator" and not have to deal with programming.

2.2 Foundations of the Displacement-Based Finite Element Method

Since it is difficult to directly solve the differential equations of linear elasticity for complex structural problems, we will consider instead work-energy principles. Let us again give the principle of virtual work (see Eq. (1.10)) here:

$$\int_V \{\delta e\}^T \{\sigma\} dV = \int_V \{\delta u\}^T \{f\} dV + \int_S \{\delta u\}^T \left\{T^{(\mathbf{n})}\right\} dS \tag{2.1}$$

Satisfying this principle is equivalent to satisfying the differential equations of equilibrium [1]. To turn this equation into a set of useful algebraic equations, let us first express the stresses in terms of the strains using the stress-strain relations (see Eq. (1.12)). Then Eq. (2.1) becomes

$$\int_V \{\delta e\}^T [C] \{e\} dV = \int_V \{\delta u\}^T \{f\} dV + \int_S \{\delta u\}^T \left\{T^{(\mathbf{n})}\right\} dS \tag{2.2}$$

Now, let the displacements and strains at a point $\mathbf{x}$ in an elastic body which occupies a volume V be represented in the forms:

$$\begin{aligned} \{u(\mathbf{x})\} &= [H(\mathbf{x})]\{U\} \\ \{e(\mathbf{x})\} &= [\mathcal{L}_E]^T \{u(\mathbf{x})\} = [\mathcal{L}_E]^T [H(\mathbf{x})]\{U\} = [B(\mathbf{x})]\{U\} \end{aligned} \tag{2.3}$$

where $[H(\mathbf{x})]$ is a matrix of "shape" functions and $\{U\}$ is a vector of displacements which are usually defined at a set of discrete points (called nodes) on the surface S of the body. The matrix $[B(\mathbf{x})]$ comes from the differentiated displacement shape functions, where $[\mathcal{L}_E]^T$ is the transpose of the differential operator matrix for the equilibrium equations defined in Eq. (1.3). This is in fact the approach used in a standard displacement-based finite element method. Placing these expressions into the principle of virtual work, Eq. (2.2), gives

$$\{\delta U\}^T \int_V [B]^T[C][B]dV\{U\} = \{\delta U\}^T \int_V [H]^T\{f\}dV + \{\delta U\}^T \int_S [H]^T\left\{T^{(\mathbf{n})}\right\}dS \quad (2.4)$$

which must be satisfied for all virtual variations of the nodal displacements, leading to a set of linear algebraic equations for those displacements:

$$[K]\{U\} = \{P\} \quad (2.5)$$

where

$$[K] = \int_V [B]^T[C][B]dV \quad (2.6)$$

is called the *stiffness matrix* and $\{P\} = \{Q\} + \{P_S\}$ where

$$\begin{aligned} \{Q\} &= \int_V [H]^T\{f\}dV \\ \{P_S\} &= \int_S [H]^T\{T^{(\mathbf{n})}\}dS \end{aligned} \quad (2.7)$$

The $\{Q\}$ and $\{P_S\}$ vectors represent discrete forces at the nodes due to any distributed body forces and surface tractions, and $\{P\}$ is just the sum of those discrete forces.

The equations of Eq. (2.5) are for a single elastic body, but if we take that body to be a small element in an otherwise complex structure, then we can in principle assemble (i.e., combine) all the elements of the structure into a very large global stiffness matrix and set of discrete displacements and forces in the same form as Eq. (2.5):

$$[K_g]\{U_g\} = \{P_g\} \quad (2.8)$$

where $[K_g]$ is the "global" stiffness matrix for the structure and $\{U_g\}$, $\{P_g\}$ are the global displacements and forces. If one applies the appropriate boundary conditions on the forces and displacements for a structure, then we can solve Eq. (2.8) for the displacements [2]. Once those discrete displacements are known, we can find the stresses in each element since we have

$$\begin{aligned} \{u(\mathbf{x})\} &= [H(\mathbf{x})]\{U\} \\ \{\sigma(\mathbf{x})\} &= [C]\{e(\mathbf{x})\} = [C][B(\mathbf{x})]\{U\} \end{aligned} \quad (2.9)$$

These steps are the essence of the finite element method in common use today. This displacement-based (or, equivalently, stiffness-based) finite element approach has become the method of choice for solving complex structural analysis problems, which in most cases are statically indeterminate. Alternative force-based approaches for statically indeterminate problems (which were described briefly in Chap. 1 as the first-generation force-based method) have largely been replaced by the displacement-based finite element approach since that first-generation force-based method, unlike the displacement-based finite element method, is not as suitable for computer-based implementation. This is unfortunate since engineers, as mentioned previously, are often most interested in forces, not displacements. However, we can reexamine work energy methods, leading to a second-generation force-based approach, where the forces are the primary unknowns.

2.3 Forces in thc Principle of Virtual Work

In contrast to the standard finite element approach, which is based solely in displacements and their derivatives, let the displacements again be defined in terms of shape-functions:

$$\{u(\mathbf{x})\} = [H(\mathbf{x})]\{U\} \tag{2.10}$$

but let the stresses be defined using separate shape functions $[N(\mathbf{x})]$, i.e.,

$$\{\sigma(\mathbf{x})\} = [N(\mathbf{x})]\{F\} \tag{2.11}$$

in terms of discrete parameters, $\{F\}$, which we will call generalized forces since, as we will see in the following chapters, they typically represent either forces or moments. The displacements $\{U\}$ will also represent generalized displacements since they may be displacements or rotations. Unlike the generalized displacements $\{U\}$, the generalized forces $\{F\}$ are not necessarily defined at the points (nodes) where the displacements act. Instead, we can consider the forces $\{F\}$ to represent parameters that describe the stresses in the volume V. In the differential equations of linear elasticity, the stresses satisfy the equilibrium equations (see Eq. (1.1)).

$$[\mathcal{L}_E]\{\sigma\} + \{f\} = \{0\} \tag{2.12}$$

The stress shape functions $[N(\mathbf{x})]$ will also be chosen so that they represent solutions of the homogeneous equilibrium equations, i.e., $[\mathcal{L}_E][N] = 0$. For axial load, torsion, and bending problems, as found in the statics and strength of materials problems considered in later chapters, such homogeneous solutions can be easily obtained. For more general two-dimensional and three-dimensional elasticity problems, one can use stress functions to define these shape functions [3, 4]. Placing these shape functions into the principle of virtual work, Eq. (2.1) then gives

$$\{\delta U\}^T \int_V [B]^T [N] dV \{F\} = \{\delta U\}^T \int_V [H]^T \{f\} dV + \{\delta U\}^T \int_S [H]^T \left\{T^{(\mathbf{n})}\right\} dS \tag{2.13}$$

where $[B] = [\mathcal{L}_E]^T [H]$ (see Eq. (2.3)). Since the variations of the virtual displacements are arbitrary, we have

$$\int_V [B]^T [N] dV \{F\} = \int_V [H]^T \{f\} dV + \int_S [H]^T \left\{T^{(\mathbf{n})}\right\} dS \tag{2.14}$$

which we will write as

$$[E]\{F\} = \{Q\} + \{P_s\} = \{P\} \tag{2.15}$$

where

$$[E] = \int_V [B]^T [N] dV \tag{2.16}$$

is an equilibrium matrix for the equilibrium equations given in Eq. (2.15) and we have

$$\begin{aligned} \{Q\} &= \int_V [H]^T \{f\} dV \\ \{P_S\} &= \int_S [H]^T \left\{T^{(\mathbf{n})}\right\} dS \end{aligned} \tag{2.17}$$

so the $\{Q\}$ and $\{P_S\}$ represent the same known discrete forces at the nodes due to distributed body forces and surface tractions found in the displacement-based finite element method, and $\{P\}$ is again just the total known force vector. A major difference with the displacement-based finite element method, however, is that even if we assemble these equilibrium relations into a global set of equilibrium equations for a complex structure in the form

$$[E_g]\{F_g\} = \{P_g\} \tag{2.18}$$

we cannot solve these equations for the forces unless the structure is statically determinate. This is not surprising since the same is true for the differential equations of elasticity where the equilibrium equations for the stresses are in general insufficient to solve for stresses. Thus, Eq. (2.18) is not by itself the basis for a general force-based finite element method. To develop such a method, we will use a combination of the principle of virtual work and complementary virtual work. This will lead us to the algebraic equations of linear elasticity defined in Chap. 1 which served as the basis for the reduced force-based and displacement-based methods described in that chapter.

2.4 Forces in the Principle of Complementary Virtual Work

We will bring into the problem the principle of complementary virtual work (see Eq. (1.11)):

$$\int_V \{\delta\sigma\}^T\{e\}dV = \int_V \{\delta f\}^T\{u\}dV + \int_S \left\{\delta T^{(\mathbf{n})}\right\}^T\{u\}dS \tag{2.19}$$

This principle is equivalent to satisfying the differential strain-displacement relations of linear elasticity [1]. The virtual changes of the stresses were assumed to satisfy

$$[\mathcal{L}_E]\{\delta\sigma(\mathbf{x})\} = [\mathcal{L}_E][N(\mathbf{x})]\{\delta F\} = \{0\} \tag{2.20}$$

However, from the differential equations, these virtual changes must also satisfy

$$[\mathcal{L}_E]\{\delta\sigma\} + \{\delta f\} = \{0\} \tag{2.21}$$

so it follows that $\{\delta f\} = 0$, and the principle of complementary virtual work can be written as

$$\int_V \{\delta\sigma\}^T\{e\}dV = \int_S \left\{\delta T^{(\mathbf{n})}\right\}\{u\}\,dS \tag{2.22}$$

Using the stress-strain relations in the form

$$\{e\} = [D]\{\sigma\} \tag{2.23}$$

where $[D] = [C]^{-1}$ is the elastic compliance matrix, and placing the shape functions into Eq. (2.22), we find:

$$\{\delta F\}^T \int_V [N]^T[D][N]dV\{F\} = \int_S \left\{\delta T^{(\mathbf{n})}\right\}^T[H]dS\{U\} \tag{2.24}$$

But since the variations of the stresses $\{\delta\sigma\}$ satisfy homogeneous equilibrium equations where $\{\delta f\} = 0$, from Eq. (2.14) we have:

$$\int_V [B]^T [N] dV \{\delta F\} = \int_S [H]^T \left\{\delta T^{(\mathbf{n})}\right\} dS \tag{2.25}$$

which we can write as

$$[E]\{\delta F\} = \int_S [H]^T \left\{\delta T^{(\mathbf{n})}\right\} dS \tag{2.26}$$

Placing the transpose of this result into Eq. (2.24) gives:

$$\{\delta F\}^T [G]\{F\} = \{\delta F\}^T [E]^T \{U\} \tag{2.27}$$

where we will call

$$[G] = \int_V [N]^T [D][N] dV \tag{2.28}$$

the flexibility matrix.

Note: The flexibility matrix [G] is different from the flexibility matrix defined in a first-generation force-based method, as discussed in Chap. 1. We could call this [G] matrix by a different name, but since we will not deal with first-generation force-based methods in this book, this distinction is not necessary.

Since the variations of the generalized forces are arbitrary, we have:

$$[G]\{F\} = [E]^T \{U\} \tag{2.29}$$

If we define a set of deformations (generalized strains), $\{\Delta\}$, as

$$\{\Delta\} = [G]\{F\} \tag{2.30}$$

then Eq. (2.29) becomes

$$\{\Delta\} = [E]^T \{U\} \tag{2.31}$$

2.5 The Algebraic Equations of Linear Elasticity

From the results of the last section, we have a complete set of equations from work-energy principles by combining Eq. (2.15), (2.30), and (2.31). They are the algebraic equations of linear elasticity, which were simply written down previously (see Eq. (1.13)):

$$\begin{aligned} [E]\{F\} &= \{P\} \\ \{\Delta\} &= [E]^T\{U\} \\ \{\Delta\} &= [G]\{F\} \end{aligned} \tag{2.32}$$

As shown in Chap. 1, these equations form the basis for the force-based method where the forces can be found from

$$\begin{bmatrix} [E] \\ - \\ [S][G] \end{bmatrix}\{F\} = \begin{Bmatrix} \{P\} \\ - \\ \{0\} \end{Bmatrix} \tag{2.33}$$

The discrete displacements $\{U\}$ do not appear directly in Eq. (2.33) but note that the displacement shape functions did enter in defining the equilibrium matrix $[E]$ and the discrete forces $\{P\}$ (see Eq. (2.16) and Eq. (2.17)). Once the forces are found from Eq. (2.33) in Chap. 1, we showed that the displacements $\{U\}$ can also be found from Eq. (1.29):

$$[E_{sys}]^T\begin{Bmatrix} \{U\} \\ - \\ \{0\} \end{Bmatrix} = [G]\{F\} \tag{2.34}$$

In this book, we will be primarily interested in the force-based method as it provides an important new path for solving statically indeterminate problems. However, as seen in Chap. 1, the algebraic equations of elasticity also lead to a displacement-based approach where the displacements can be found from

$$[E][G]^{-1}[E]^T\{U\} = \{P\} \tag{2.35a}$$

or equivalently,

$$[K_p]\{U\} = \{P\} \tag{2.35b}$$

in terms of the pseudo stiffness matrix $[K_p]$.

The forces can then be found from

$$\{F\} = [G]^{-1}[E]^T\{U\} \tag{2.36}$$

This is a displacement-based approach because only the displacements appear in Eqs. (2.35a and 2.35b), but notice that the pseudo stiffness matrix does depend on the stress shape functions present in both $[E]$ and $[G]$. The traditional displacement-based finite element only uses the principle of virtual work and the displacement shape functions or their derivatives so that it is truly based only on displacements.

When determining displacements, there also is a rather subtle point regarding signs. In the traditional displacement-based finite element method, the force vector $\{P\}$ is normally assembled from the applied forces acting at the nodes in the positive (x, y, z) directions, i.e., $\{P\} = \{P\}_{x,\,y,\,z}$, so the displacements solved for are also the displacements in the positive (x, y, z) directions, i.e.,

$$[K]\{U\}_{x,y,z} = \{P\}_{x,y,z} \tag{2.37}$$

The same result is true if we use the pseudo stiffness matrix and assemble the force vector in the same way, i.e.,

$$\left[K_p\right]\{U\}_{x,y,z} = \{P\}_{x,y,z} \tag{2.38}$$

However, in formulating the equilibrium equations from free body diagrams at the nodes, it is common practice to relate the unknown forces $\{F\}$ and the external applied loads, $\{P\}$ by requiring that the sum of all the forces and moments in the positive (x, y, z) directions to be zero, i.e.,

$$\begin{matrix} \sum\limits_{x,y,z} \mathbf{F} = 0 \\ \sum\limits_{x,y,z} \mathbf{M} = 0 \end{matrix} \rightarrow [E]\{F\}_{x,y,z} + \{P\}_{x,y,z} = 0 \tag{2.39}$$

If we place all the applied load terms on the right-hand side of these equations, we obtain

$$[E]\{F\}_{x,y,z} = -\{P\}_{x,y,z} \tag{2.40}$$

Thus, we see that when using the equilibrium equations of Eq. (2.39) to write the vector-matrix equilibrium equations as $[E]\{F\} = \{P\}$, which we will consistently do in this book when forming the equilibrium equations on a node-by-node basis, we have

$$[E]\{F\} = \{P\} = -\{P\}_{x,y,z} \tag{2.41}$$

Using this value for $\{P\}$ with either the stiffness matrix or pseudo stiffness matrix, we find

$$
\begin{aligned}
[K]\{-U\}_{x,y,z} &= -\{P\}_{x,y,z} \\
[K_p]\{-U\}_{x,y,z} &= -\{P\}_{x,y,z}
\end{aligned}
\tag{2.42}
$$

and the displacements we solve for will be the negative of the actual displacements $\{U\}_{x,\,y,\,z}$ acting in the positive (*x*, *y*, *z*) directions. If the same value for $\{P\}$, namely $\{P\} = -\{P\}_{x,\,y,\,z}$, is used in defining the equilibrium equations when solving for the unknown forces $\{F\}$ in the second-generation force-based method, we will also obtain the negative of the displacements when solving for the displacements, i.e.,

$$
\begin{Bmatrix} \{-U\}_{x,y,z} \\ - \\ \{0\} \end{Bmatrix} = \mathrm{inv}\left(\left[E_{sys}\right]^T\right)[G]\{F\}
\tag{2.43}
$$

We will see these negative signs appearing in the displacements when we consider solutions for specific problems.

Note: We could choose to multiply both the equilibrium matrix $[E]$ and the force vector $\{P\}$ by minus signs since that will not change the solution of the equilibrium equations but it will make the vector $\{P\}$ be the forces/moments in the positive (*x*, *y*, *z*) directions. If we make those changes, no minus signs will appear when determining the displacements. We will not adopt that approach in Chaps. 3, 4 and 5, where we will form the equilibrium equations on a node-by-node basis. However, we will see in Chap. 6, in a force-based finite element method where the equilibrium equations are generated on an element-by-element basis, the element equilibrium matrix used will be the negative of the one formed on a node-by-node basis and the vector $\{P\}$ will be the forces/moments in the positive (*x*, *y*, *z*) directions so that no minus signs will appear when finding displacements.

There is another point we want to emphasize about displacements. In the algebraic equations of linear elasticity, the number of displacements is equal to the number of equilibrium equations. For example, if we have a force equilibrium equation in a given direction, there will be a linear displacement, U, acting in the same direction associated with that equation. Similarly, if we have a moment equilibrium equation in a given direction, taken about a point A, there will be an angular displacement, θ_A, in that same direction about A associated with that equation. Thus, for example, if we have three equilibrium force equations and three equilibrium moments equations in the (*x*, *y*, *z*) directions, these equations will be associated with six displacements:

$$
\begin{array}{lcl}
\sum F_x = 0 & \leftrightarrow & U_x \\
\sum F_y = 0 & \leftrightarrow & U_y \\
\sum F_z = 0 & \leftrightarrow & U_z \\
\sum M_{xA} = 0 & \leftrightarrow & \theta_{xA} \\
\sum M_{yA} = 0 & \leftrightarrow & \theta_{yA} \\
\sum M_{zA} = 0 & \leftrightarrow & \theta_{zA}
\end{array}
\tag{2.44}
$$

The displacements found will be the negative of those appearing in Eq. (2.44) if the equilibrium equations are written as $[E]\{F\} = \{P\}$, where $\{P\} = -\{P\}_{x,\,y,\,z}$, as just discussed, and the physical meaning of these displacements will depend on the nature of the problem being considered, as we will see in examples.

2.6 Solving Statically Indeterminate Problems with MATLAB

2.6.1 *Force-Based Solutions*

Problems that are defined by the algebraic equations of linear elasticity can be solved for the forces with the force-based method or for the displacements with a pseudo stiffness matrix approach, provided that the assumptions used in the derivation of those algebraic equations are satisfied. They are:

1. The applied generalized forces $\{P\}$ in the equilibrium equations $[E]\{F\} = \{P\}$ must be purely discrete quantities so that any distributed forces (i.e., forces that are functions of location) must be replaced by equivalent discrete loads. The $\{P\}$ are generally known generalized forces, but discrete unknown forces (usually at supports) can also be considered as part of $\{P\}$.
2. The unknown internal forces $\{F\}$ were assumed to be discrete quantities that represent the internal stresses (or equivalently the internal generalized forces) in the elastic parts of a structure and that satisfy homogeneous equilibrium equations in those parts. However, we will also be able to consider problems where $\{F\}$ can include unknown reaction forces or forces in rigid parts of a system.
3. The flexibilities of the elastic parts must be specified. Generally, these are at most small square matrices. These must be placed in a flexibility matrix $[G]$ that defines the relationship between the deformations $\{\Delta\}$ and the internal

(continued)

forces $\{F\}$ in the form $\{\Delta\} = [G]\{F\}$. The fact that this is a linear relationship means that we are assuming the elastic parts are acting as *linear elastic* bodies. It will be shown that when $\{F\}$ includes unknown forces associated with rigid connections or supports, those can also be included in the flexibility matrix by setting their flexibilities to zero.

4. The displacements and deformations of the elastic parts are very small, so that when drawing and using free body diagrams to obtain the equilibrium equations, they can be ignored. Assuming the deformations are small also means that the assumption of linear elastic behavior is reasonable since many structures behave in this manner if the deformations are not large.

If these four requirements are met, then it is possible to use algebraic equations to solve for either the forces or for the displacements directly. MATLAB has many built-in functions that perform all the necessary calculations. The creation of the equilibrium matrix is perhaps the most complex task. In a finite element approach, this can be done by assembling either small pseudo stiffness matrices or small equilibrium matrices for individual elements into a global matrix for the entire structure with the use of computers and programming. This is the most efficient way to proceed when dealing with complex problems. In many of the problems considered in an engineering curriculum, however, the number of equations of equilibrium is not large and they can be written explicitly on paper. In this case, the equilibrium equations can also be written in MATLAB in symbolic form. The built-in symbolic and numerical tools and functions of MATLAB can then be used to carry out all needed steps and obtain the solution *without any programming*. This makes MATLAB an ideal tool for solving statically indeterminate problems at many levels in an engineering curriculum with the algebraic methods described here.

Chapters 3, 4, and 5 will demonstrate this use of MATLAB for the typical problems found in statics, strength of materials, and advanced strength of materials, respectively. Here, a simple example (and some variations of it) will be given to illustrate the steps involved in obtaining a solution and allow us to discuss the types of issues that arise in the solution process.

Example 2.1: A Rigid Beam on a Flexible Foundation

Consider the case shown in Fig. 2.1a where a rigid beam rests on a flexible foundation that is modeled by eleven equally spaced linear springs, each having a spring constant $k = 1000$ lb/ft. A 400 lb force acts on the beam, whose own weight can be neglected. Determine the forces in the springs and plot those forces along the length of the beam. Also, determine the displacements.

(continued)

Example 2.1 (continued)

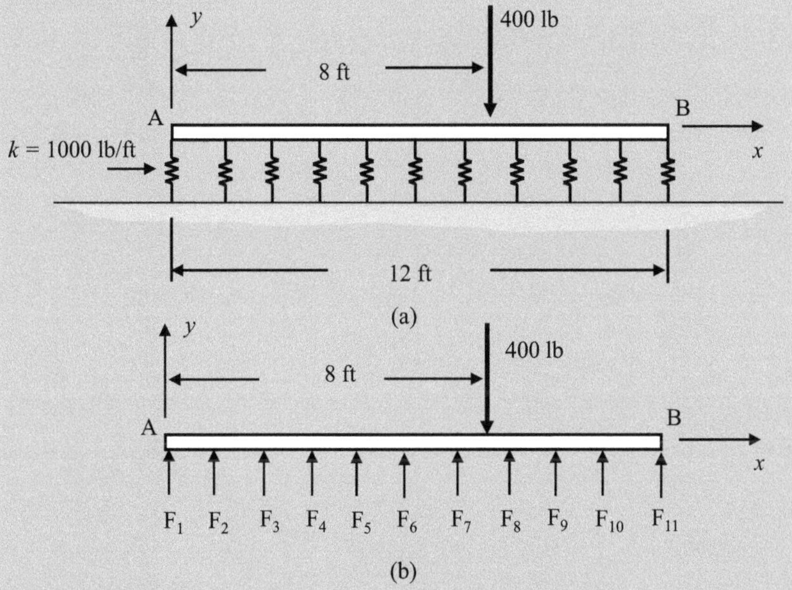

Fig. 2.1 (**a**) A beam resting on a deformable foundation that is modeled as a distribution of springs. (**b**) A free body diagram of the beam

From the free body diagram of Fig. 2.1b, summing forces in the y-direction and moments about A in the z-direction, the equations of equilibrium are

$$\begin{aligned}
&\sum F_y = 0 \quad F_1 + F_2 + F_3 + F_4 + F_5 + F_6 + F_7 + F_8 \\
&\quad + F_9 + F_{10} + F_{11} - 400 = 0 \\
&\sum M_{zA} = 0 \quad 1.2F_2 + 2.4F_3 + 3.6F_4 + 4.8F_5 + 6F_6 \\
&\quad + 7.2F_7 + 8.4F_8 + 9.6F_9 + 10.8F_{10} + 12F_{11} - (400)(8) = 0
\end{aligned} \tag{2.45}$$

which are two equations for eleven unknown forces, so the problem is highly statically indeterminate. Since the internal force F in a linear spring is related to the spring deformation (elongation) Δ through $\Delta = F/k$, the flexibility of the spring is $g = 1/k$, i.e., it is the reciprocal of the spring constant k. Note that an internal force in a spring is just a constant F anywhere in the spring (see Fig. 2.2a) so that the spring forces appearing in Eq. (2.45) do satisfy homogeneous equilibrium equations, i.e. they are in equilibrium with themselves. The same is true for the constant force in a two-force elastic member such as a wire or a truss member (Fig. 2.2b), structures that are often considered in statics. All the forces acting in the springs in this problem are discrete compressive forces and that is consistent with the spring forces being shown as pushing on the beam in Fig. 2.1b. Now, let's solve this problem in MATLAB.

(continued)

Example 2.1 (continued)

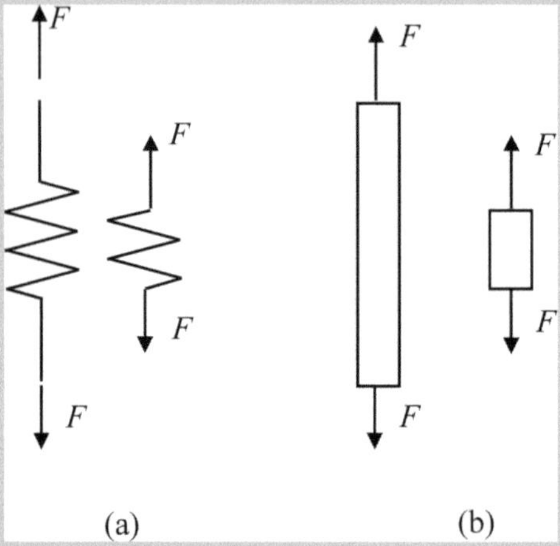

Fig. 2.2 (**a**) A spring in equilibrium carries a force F that is constant for any part of the spring. (**b**) A two-force member in equilibrium carries a force F that is constant in the member. In both cases, F is shown as a tensile force

First, define the variables symbolically that appear in the equilibrium equations

```
syms F1 F2 F3 F4 F5 F6 F7 F8 F9 F10 F11 % the unknown forces
```

Then, write symbolically the two equilibrium equations

```
Eq(1) = F1 + F2 + F3 + F4 + F5 + F6 + F7 + F8 + F9 + F10 + F11 -400;
Eq(2) = 1.2*F2 + 2.4*F3 + 3.6*F4 + 4.8*F5 + 6.0*F6 + 7.2*F7 + ...
         8.4*F8 + 9.6*F9 + 10.8*F10 + 12*F11 - (400)*(8);
```

Next, place all the unknown forces in a vector

```
T = [ F1 F2 F3 F4 F5 F6 F7 F8 F9 F10 F11];
```

and use the MATLAB function equationsToMatrix to obtain the equilibrium matrix E = $[E]$ and the "force" vector P = $\{P\}$

```
[E, P] = equationsToMatrix(Eq, T);
```

The MATLAB null function can be used to obtain the solutions of the homogeneous equilibrium equations. Taking the matrix transpose then gives us the compatibility matrix S = $[S]$

```
S = null(E).';
```

(continued)

Example 2.1 (continued)

Specify the spring constant, which is the same for all springs

```
k = 1000;
```

and form up a vector with all eleven flexibilities

```
flex = (1/k)*ones(1,11);
```

The MATLAB function diag then places all these flexibilities along the diagonal of a flexibility matrix, G = [G]

```
G = diag(flex);
```

We now can form up the system equations [Esys]{F} = {Psys} by combining our equilibrium and compatibility equations (there are nine compatibility equations, so we need nine zeros added in a column to the column vector {P})

```
Esys = [ E; S*G];
Psys = [P; zeros(9,1)];
```

and we can solve this system of equations with the MATLAB solver \ . The solution will still be a symbolic vector, so we also convert the solution to numerical double precision values with the MATLAB double function

```
F = double(Esys\Psys); % solve and convert to double precision
```

The forces F1 to F11 are:

```
F % the forces in the springs (in lb)
F =

 6.0606
 12.1212
 18.1818
 24.2424
 30.3030
 36.3636
 42.4242
 48.4848
 54.5455
 60.6061
 66.6667
```

(continued)

Example 2.1 (continued)

Finally, we use a set of eleven values to plot these eleven forces with the MATLAB plot function

```
x = 1:1:11; % plot the forces
plot( x, F, 'o')
xlabel('spring number')
ylabel(' spring force (lb) ')
```

The plot of the spring forces is shown in Fig. 2.3. Now consider the displacements in this problem. The transpose of the equilibrium matrix is a 6×2 matrix, so we see from the deformation-displacement relation $\{\Delta\} = [E]^T\{U\}$, there are two displacements $\{U\} = [u \ \ \theta]^T$. In this problem, we will show that when we solve for these displacements, we will obtain the translational displacement, u, of the rigid beam and the rotational angle, θ, about A it makes with the x-axis as shown in Fig. 2.4. Note that $u = -u_y$, $\theta = -\theta_z$, so both u and θ are the negative of the displacements in the positive y- and z-directions, consistent with our previous discussions. Now, let's use MATLAB to obtain these deformation-displacement relations:

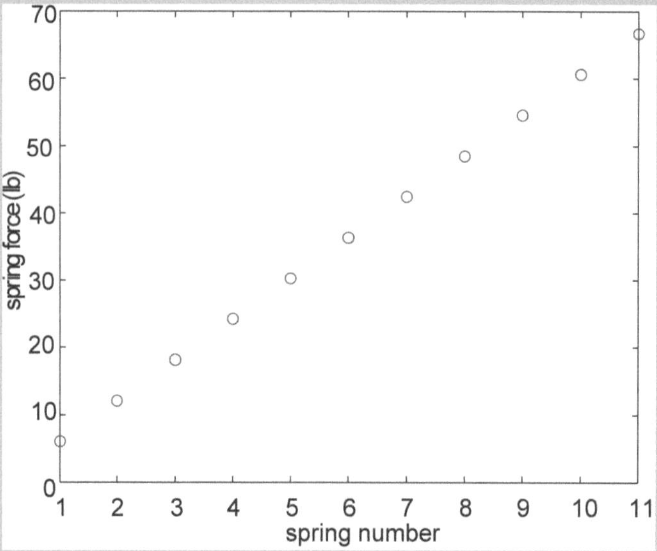

Fig. 2.3 A plot of the spring forces supporting the beam

(continued)

Example 2.1 (continued)

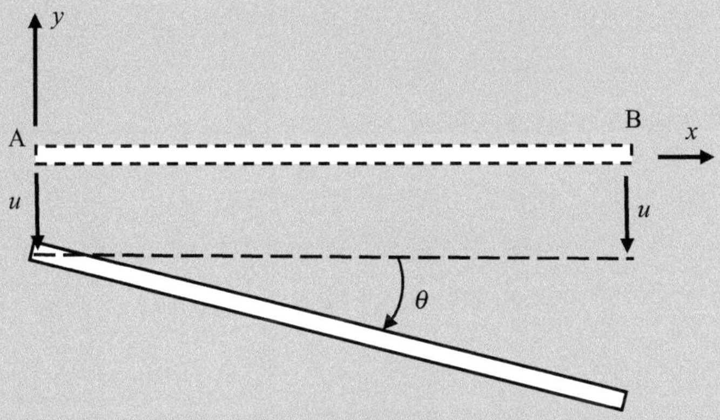

Fig. 2.4 The displacements of the rigid beam

First, define symbolically the displacement u and angle theta = t symbolically

```
syms u t
```

Then place these displacements in a column vector {U}

```
U= [u;t];
```

Now, compute the deformations $\{D\} \equiv \{\Delta\}$, where $\{D\} = [E]^T\{U\}$

```
D=E.'*U
```

```
D=u
 (6*t)/5 + u
 (12*t)/5 + u
 (18*t)/5 + u
 (24*t)/5 + u
 6*t + u
 (36*t)/5 + u
 (42*t)/5 + u
 (48*t)/5 + u
 (54*t)/5 + u
 12*t + u
```

(continued)

Example 2.1 (continued)

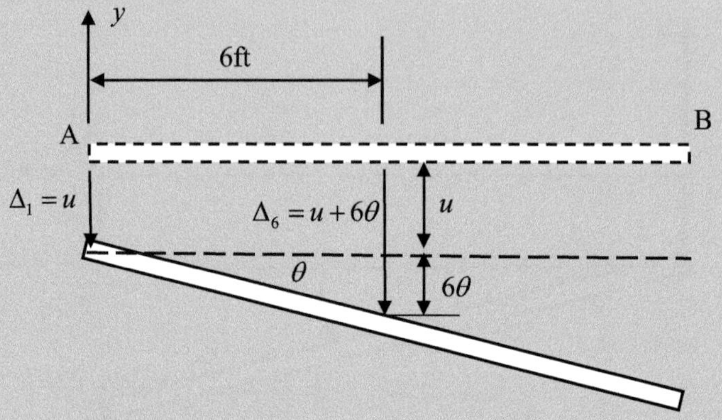

Fig. 2.5 The relationship between the deformation and the displacement and rotation of the beam for the spring at A and the spring at the middle of the beam

If we consider, say, the deformation of the spring at A and the spring at the middle of the beam, we see from Fig. 2.5 that we have (for small displacements and rotations) $\Delta_1 = u$ and $\Delta_6 = u + 6\theta$ which agree with the result from the deformation-displacement relation $\{\Delta\} = [E]^T\{U\}$. The other deformations and displacements are likewise obtained from the geometry changes in Fig 2.1b. If we calculate these displacements from the forces we find

```
Usys = Esys.'\G*F;
double(Usys)
Usys =
 0.0061
 0.0051
 0
 0
 0
 0
 0
 0
 0
 0
 0
```

so we have the two displacements $u = .0061$ ft and $\theta = 0.0051$ radians and nine additional zeros. Now, consider more explicitly the compatibility

(continued)

Example 2.1 (continued)
equations in this problem. First, define symbolically the eleven deformations, which we will call D1 to D11

```
syms D1 D2 D3 D4 D5 D6 D7 D8 D9 D10 D11
```

and place them in a column vector DV where DV = $\{\Delta\}$

```
DV = [D1 D2 D3 D4 D5 D6 D7 D8 D9 D10 D11].';
```

Next, examine explicitly the compatibility matrix [S] we determined previously

```
S
S = [1, -2, 1, 0, 0, 0, 0, 0, 0, 0, 0]
 [2, -3, 0, 1, 0, 0, 0, 0, 0, 0, 0]
 [3, -4, 0, 0, 1, 0, 0, 0, 0, 0, 0]
 [4, -5, 0, 0, 0, 1, 0, 0, 0, 0, 0]
 [5, -6, 0, 0, 0, 0, 1, 0, 0, 0, 0]
 [6, -7, 0, 0, 0, 0, 0, 1, 0, 0, 0]
 [7, -8, 0, 0, 0, 0, 0, 0, 1, 0, 0]
 [8, -9, 0, 0, 0, 0, 0, 0, 0, 1, 0]
 [9, -10, 0, 0, 0, 0, 0, 0, 0, 0, 1]
```

This matrix should satisfy $[S]\{\Delta\} = \{0\}$, so let's see if it does. Let's compute $[S]\{\Delta\}$:

```
S*DV
 ans =
 D1 - 2*D2 + D3
 2*D1 - 3*D2 + D4
 3*D1 - 4*D2 + D5
 4*D1 - 5*D2 + D6
 5*D1 - 6*D2 + D7
 6*D1 - 7*D2 + D8
 7*D1 - 8*D2 + D9
 8*D1 - 9*D2 + D10
 9*D1 - 10*D2 + D11
```

(continued)

Example 2.1 (continued)

The first row in the result above says that the deformations must be related by $\Delta_1 - 2\Delta_2 + \Delta_3 = 0$, and from the deformation-displacement relations we found $\Delta_1 = u$, $\Delta_2 = 6\theta/5 + u$, $\Delta_3 = 12\theta/5 + u$; so we do find

$$u - 2(6\theta/5 + u) + (12\theta/5 + u) = 0 \tag{2.46}$$

and this first compatibility equation for the deformations is indeed satisfied independent of what the displacements may be. The other compatibility relations can be shown to be satisfied in the same fashion. We could have in principle determined these nine compatibility equations by looking at the geometry for all eleven springs but that would be rather difficult and tedious in comparison to simply determining them from homogeneous solutions of the equilibrium matrix with the MATLAB null function.

In this problem, the flexibility matrix [G] was a diagonal matrix with the flexibilities 1/k along the diagonal. Since the flexibilities were all the same, when solving for the forces, we could have removed the common 1/k values and still have found the forces properly. This is because the compatibility equations $[S][G]\{F\} = \{0\}$ are homogeneous equations so we can always multiply them by an arbitrary constant value. However, *when solving for the displacements, the flexibilities must be present, and the units used in the equilibrium matrix and the flexibility matrix must be consistent.* They were consistent here since the force and moment equations were in terms of lb and ft-lb and the flexibilities were given in terms of ft/lb.

2.6.2 Pseudo Stiffness Matrix

Although our primary focus in this book is on force-based methods, we could also solve this problem with the use of a pseudo stiffness matrix. Here are the steps in MATLAB.

Since the equilibrium matrix $[E]$ and the flexibility matrix $[G]$ have already been obtained, we can form the pseudo stiffness matrix:

```
Kp =E*inv(G)*E.';
```

The applied force vector $\{P\}$ has also been obtained previously so we can solve for the displacements

```
U = Kp\P;
Un = double(U)
Un = 0.0061
     0.0051
```

which agrees with our previous results. Once the displacements are known, then the forces are given by $\{F\} = [G]^{-1}[E]^T\{U\}$:

```
F = double(inv(G)*E.'*U)
F = 6.0606
   12.1212
   18.1818
   24.2424
   30.3030
   36.3636
   42.4242
   48.4848
   54.5455
   60.6061
   66.6667
```

which also agrees with our previous results.

Example 2.2: Beam on a Flexible Foundation and Pinned at an End
Including Forces at Fixed Supports

In the beam problem of Fig. 2.1, all the unknown forces were for elastic springs. Suppose, however, the beam was pinned at A as shown in Fig. 2.6a. In that case, the force at A will be a reaction force A_y and there will be no deformation or displacement at that point. The free body diagram of Fig. 2.6b looks identical to the one in Fig. 2.1 except now the spring force F_1 is replaced by A_y. Thus, the two equilibrium equations will be the same as before when we make that replacement. Since there are two equilibrium equations, we expect to again have two displacements (u, θ), but physically we expect the beam can now only rotate about A (see Fig. 2.7) so that we should find $u = 0$. We can solve this problem using exactly the same steps as before by simply modifying the flexibility of the first spring to be zero. When we solve for the forces, we will then find $F_1 = A_y$ and the other ten spring forces. Here is the solution in MATLAB, assuming we have generated $[E]$, $\{P\}$, $[S]$, and $[G]$ as before:

Set the flexibility of the first spring to be zero in the flexibility matrix

```
G(1,1) = 0;
```

Form up and solve the system equations for the forces

```
Esys = [E; S*G];
F = Esys\Psys;
```

(continued)

Example 2.2 (continued)

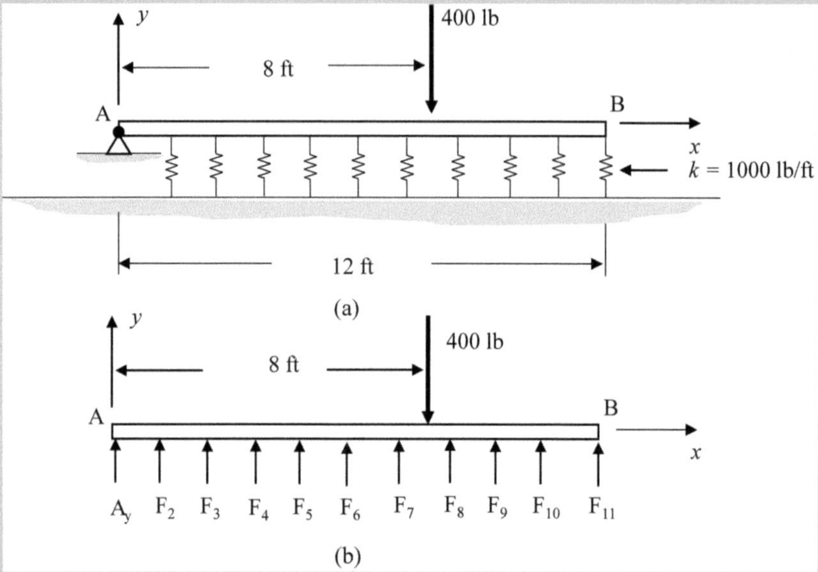

Fig. 2.6 (**a**) A rigid beam that is pinned at A and supported by springs. (**b**) The free body diagram of the beam

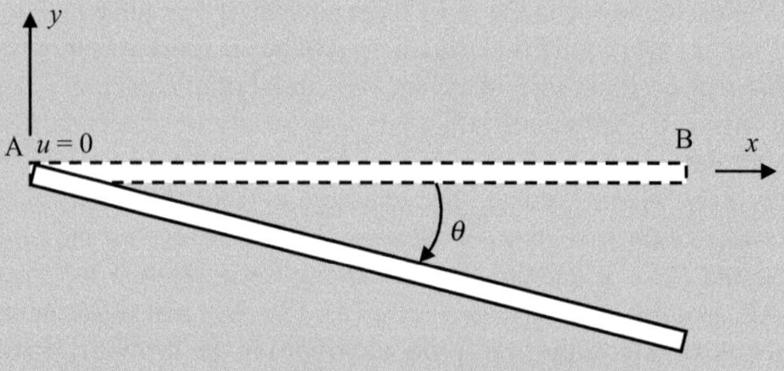

Fig. 2.7 The displacement of the rigid beam when it is pinned at A

Here are the forces. Note that the first force, which is A_y, is much larger than before because the reaction force must prevent any displacement at A.

```
double(F)
ans = 19.0476
        6.9264
```

(continued)

Example 2.2 (continued)

```
    13.8528
    20.7792
    27.7056
    34.6320
    41.5584
    48.4848
    55.4113
    62.3377
    69.2641
```

Now compute the displacements

```
u = Esys.'\G*F;
```

There are still two displacements here and nine zeros but now the first displacement $u = 0$.

```
double(u)
ans = 0
    0.0058
    0
    0
    0
    0
    0
    0
    0
    0
    0
```

Including reaction forces at fixed supports or other forces in a structure that are associated with elements of the structure where both the deformations and displacements are zero, therefore, is very easy with the force-based method. We merely need to set the flexibilities associated with those forces to be zero. We can then solve for the forces. When we solve for the displacements, we will also obtain zero values at the fixed supports. However, there are cases where forces in a structure have no deformations but where non-zero rigid body displacements are present at the locations where those forces act. In that case, when we set the flexibilities associated with those forces to zero, the associated displacements will be non-zero. We will give an example of such a special case later.

Example 2.3: Beam on a Flexible Foundation and Pinned at the Center
We can consider the problem of Fig. 2.1 in the case where we fix the beam at locations other than the end A. Suppose, for example, we want to have the beam fixed at the center of the beam at point C where the spring force F_6 acts (Fig. 2.8a). Again the free body diagram is the same except now $F_6 = C_y$, where C_y is the reaction force at that point. In this case, we can reuse the previous MATLAB solution by restoring the flexibility of the first spring back to its original value and then setting the flexibility of the middle spring to be zero to obtain a flexibility matrix for this problem. Here is the solution in MATLAB, assuming we have generated [E], {P}, and [S] for the problem of Fig. 2.1a:

First, restore the flexibility of the first spring and then set the flexibility of the middle spring to be zero.

```
G(1,1) = 1/1000;
G(6, 6) = 0;
```

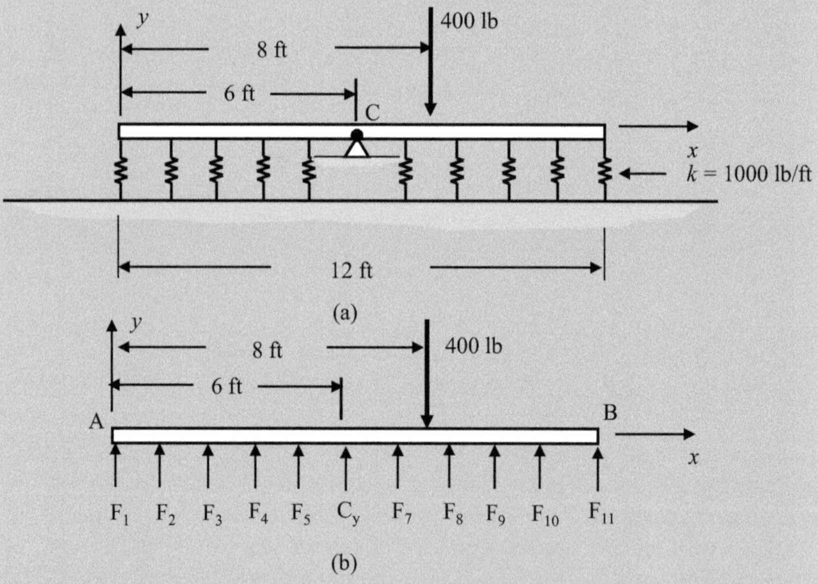

Fig. 2.8 (**a**) A rigid beam that is pinned at its center C and supported by springs. (**b**) The free body diagram of the beam

(continued)

Example 2.3 (continued)
Set up and solve the system equations for the reaction force at C, C_y, and the spring forces

```
Esys = [ E; S*G];
F = Esys\Psys;
double(F)
ans = -30.3030
      -24.2424
      -18.1818
      -12.1212
      -6.0606
      400.0000 % This is the force Cy
        6.0606
       12.1212
       18.1818
       24.2424
       30.3030
```

Now the first five springs are in tension and the last five springs are in compression. The spring forces are symmetrical in their magnitudes because of the symmetry of the deformations (Fig. 2.9). The center support force, C_y, is much larger than before because of the constraint on the beam at that point. It is equal to the applied force because the spring forces occur in canceling pairs.

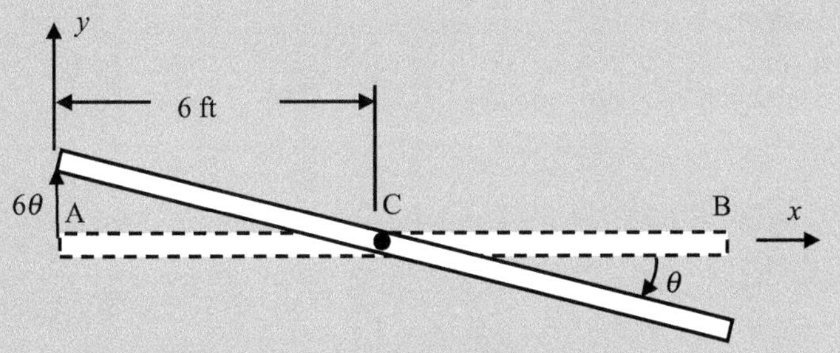

Fig. 2.9 The rotation of the rigid beam and the displacement at A when it is pinned at C

(continued)

Example 2.3 (continued)

Now, compute the displacements. Again, there are only two displacements:

```
u = Esys.'\G*F;
double(u)
ans = -0.0303
       0.0051
       0
       0
       0
       0
       0
       0
       0
       0
       0
```

The angle θ is the same as before. The displacement u is still the displacement at A because we took moments about point A for the moment equation. It is negative since u was defined to be positive downward. Since the beam is rotating about C from the geometry (Fig. 2.8), we expect the displacement at A to be $u = -6\theta$ ft. We see that is indeed the case:

```
double(-u(2)*6)
ans = -0.0303
```

If we replace the moment equation about point A with a moment equation about C, then the displacement u will be the downward displacement at C instead, which is zero. We will not show that result explicitly here.

2.6.3 Zero Flexibilities and the Pseudo Stiffness Matrix

If we set the flexibilities at fixed, rigid supports to be zero, then we cannot solve for the pseudo stiffness matrix $[K_p] = [E][G]^{-1}[E]^T$ because the inverse of the flexibility matrix does not exist. There is, however, a simple work-around to this problem. If we replace the zero flexibilities with very small flexibilities (small relative to the other flexibilities which are normally already small), then we can invert the flexibility matrix and solve approximately for the displacements. Let us demonstrate this with the last problem considered where the spring-supported beam is fixed at C. First, let us examine the displacements just obtained to a larger number of decimal places by displaying those displacements with the long format in MATLAB:

```
format long
double(u)

ans =

 -0.030303030303030
  0.005050505050505
  0
  0
  0
  0
  0
  0
  0
  0
  0
```

Now, let us modify the flexibility for the force at C so it is not exactly zero. Then we can form up a pseudo stiffness matrix and solve for the two displacements with that matrix:

```
G(6,6) = 10^-9;
Kp = E*inv(G)*E.';
u = Kp\P;
double(u)

ans =

 -0.030302630307030
  0.005050505050505
```

so we have obtained the correct displacements to about six decimal places.

This method can also show us why, when we set the flexibility associated with a force at a fixed support to be zero, the associated displacement is also zero at that support. Consider, for example, the problem of Fig. 2.6 again where the beam is pinned at A. The two equilibrium equations were the force equation in the y-direction and a moment equation about point A in the z-direction. We will make the first flexibility at the pinned support A much smaller than all the others, which we must first restore to the common value of 1/000 and then set:

```
G(1,1) = (10^-9)*G(1,1);
```

Then we will form up the pseudo stiffness matrix.

```
Kp = E*inv(G)*E.'
Kp = [1000000010000, 66000]
     [66000, 554400]
```

The stiffness Kp(1,1) is much larger than the others, so in the equilibrium equations for the displacement at A and the rotation of the beam given by

$$[K_p]\left\{\begin{array}{c} u \\ \theta \end{array}\right\} = \{P\}$$

the first equation will be approximately $u = 0$, where u in this case is the displacement at A. We can verify this by solving for the displacements:

```
U = Kp\P;
double(U)
ans = 0.0000
      0.0058
```

which gives the same displacements found previously, at least to this displayed number of decimal points.

Satisfying zero displacements (approximately) in the traditional stiffness-based finite element method is done in a very similar fashion where the stiffness matrix is modified directly with large stiffnesses to approximately satisfy zero displacement boundary conditions at fixed nodes [2].

2.6.4 *Zero Flexibilities and Rigid Body Displacements*

Examples 2.1, 2.2 and 2.3 have covered many of the important points associated with the use of a second-generation force-based method. However, we mentioned previously that satisfying the deformations to be zero at a connection does not always require the displacements there to also be zero. We will see such a case in the next example.

Example 2.4: A Frame with Springs
The X-frame shown in Fig. 2.10 is a pinned structure where the rigid members are supported by a pin and a roller and held in the configuration shown by two springs that are assumed to be very stiff. Let the springs constants have the same value of k lb/ft. The lengths of CE and CD are 5 ft and the lengths of AC and CB are 10 ft, as shown. We want to (a) find the tensions in the springs and (b) determine all the forces present in the structure. We will also discuss the displacements in this problem.

(continued)

Example 2.4 (continued)

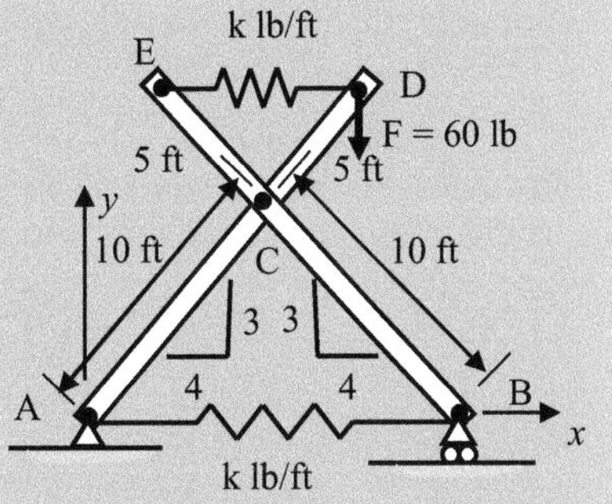

Fig. 2.10 A statically indeterminate frame problem

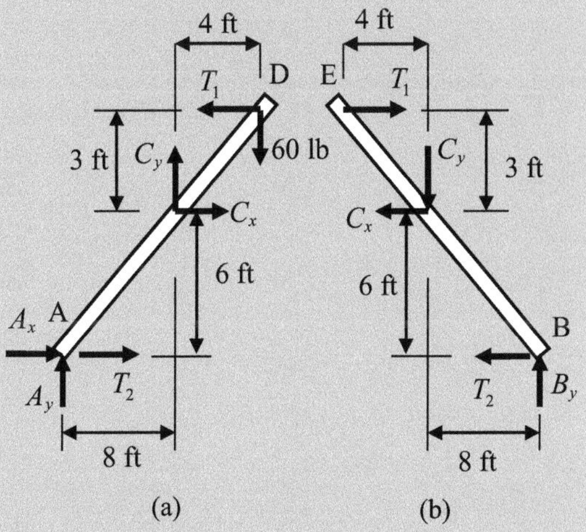

Fig. 2.11 Free body diagrams of (**a**) member AD, (**b**) member BE

The free body diagrams of the two members of the frame are shown in Fig. 2.11. The forces in the springs ED and AB are called T_1 and T_2 (note that as drawn the force in ED is assumed to be in compression and the force in AB in tension). All the other forces are due to supports or connections of the rigid members. The equilibrium equations for the two members are:

$$
\begin{aligned}
&{}_{ACD}\sum F_x = 0 \quad A_x + C_x + T_2 - T_1 = 0\\
&{}_{ACD}\sum F_y = 0 \quad A_y + C_y - 60 = 0\\
&{}_{ACD}\sum M_{zC} = 0 \quad 3T_1 - (4)(60) + 6A_x + 6T_2 - 8A_y = 0\\
&{}_{BCE}\sum F_x = 0 \quad T_1 - C_x - T_2 = 0\\
&{}_{BCE}\sum F_y = 0 \quad B_y - C_y = 0\\
&{}_{BCE}\sum M_{zC} = 0 \quad 8B_y - 6T_2 - 3T_1 = 0
\end{aligned}
\tag{2.47}
$$

(continued)

Example 2.4 (continued)

By placing these equilibrium equations symbolically into MATLAB, we can extract the equilibrium matrix and the vector of known loads with the equationsToMatrix function:

```
syms Ax Ay By Cx Cy T1 T2 k  % symbolic variables
Eq(1) = Ax + Cx + T2 - T1;   % equations of equilibrium
Eq(2) = Ay + Cy -60;
Eq(3) = 3*T1 -240 + 6*Ax + 6*T2 - 8*Ay ;
Eq(4) = T1 - Cx - T2;
Eq(5) = By - Cy;
Eq(6) = 8*By -6*T2 - 3*T1;
T = [ T1 T2 Ax Ay Cx Cy By] ;  % vector of unknown forces
[E, P] = equationsToMatrix(Eq, T) ; % the equilibrium matrix and
         load vector
```

Then, we can generate the compatibility matrix and the flexibility matrix. The flexibilities of the springs are 1/*k* and all the other flexibilities are zero:

```
S = null(E).';               % determine the compatibility matrix
flex = [ 1/k 1/k 0 0 0 0 0] ; % flexibility matrix with zero
                               flexibilities
G = diag(flex) ;
```

Next, we combine the equilibrium and compatibility equations into a set of system of equations that we can solve:

```
Esys = [ E; S*G] ; % form up the system of equations and solve
Psys = [P; 0] ;
F = Esys\Psys;
F = 24   % T1 = 24 lb
    48   % T2 = 48 lb
    0    % Ax = 0
    15   % Ay = 15 lb
   -24   % Cx = -24 lb
    45   % Cy = 45 lb
    45   % By = 45 lb
```

These results are independent of the spring constant since it could be eliminated before we solved for the forces. Now let us solve for the displacements. We find:

(continued)

Example 2.4 (continued)

```
U = Esys.'\G*F
 U =

-24/k % - Ux translational displacement of ACD
 32/k  % -Uy translational displacement of ACD
 4/k   % -angle of rotation in the z-direction of ACD about C
-24/k % -Ux translational displacement of BCE
 32/k  % -Uy translational displacement of BCE
 -4/k  % -angle of rotation in the z-direction of BCE about C
 0 % additional zero
```

The first three terms are the negative of the translational displacements and rotation (about C) of rigid member ACD, while the next three terms represent the negative of the translational displacements and rotation of rigid member BCE. To obtain the displacement of any point in a rigid body, we need to add the translational displacements to the rotational displacements from the point about which the rotations are taken to the point where we want to calculate the displacement. For example, to calculate the displacement of point A, we have

$$\mathbf{u}_A = \mathbf{u}_T^{ACD} + \left(\theta_{Cz}^{ACD}\mathbf{e}_z\right) \times \mathbf{r}_{A/C} \tag{2.48}$$

where $\mathbf{u}_T^{ACD}$ is the translational displacement of ACD, θ_{Cz}^{ACD} is the small rotation of ACD about the z-axis, and $\mathbf{u}_A$ is the displacement of point A. The vector $\mathbf{r}_{A/C}$ is the position vector from C to A. From the displacements we found for ACD (and changing their signs), we have

$$\begin{aligned}\mathbf{u}_A &= (24/k)\mathbf{e}_x - (32/k)\mathbf{e}_y + (-4/k)\mathbf{e}_z \times \left(-8\mathbf{e}_x - 6\mathbf{e}_y\right) \\ &= 0\end{aligned} \tag{2.49}$$

where ($\mathbf{e}_x$, $\mathbf{e}_y$, $\mathbf{e}_z$) are unit vectors along the (x, y, z) axes, respectively. Thus, the displacement at the pin A is indeed zero. Similarly, the displacement at B for BCE is:

$$\begin{aligned}\mathbf{u}_B &= \mathbf{u}_T^{BCE} + \left(\theta_{Cz}^{BCE}\mathbf{e}_z\right) \times \mathbf{r}_{B/C} \\ &= (24/k)\mathbf{e}_x - (32/k)\mathbf{e}_y + (4/k)\mathbf{e}_z \times \left(8\mathbf{e}_x - 6\mathbf{e}_y\right) \\ &= (48/k)\mathbf{e}_x\end{aligned} \tag{2.50}$$

which also satisfies the constraint that there is no y-displacement at B, although B can displace in the x-direction. Thus, by setting the flexibilities to be zero

(continued)

Example 2.4 (continued)
associated with the forces $(A_x, A_y, B_y, C_x, C_y)$, we did also satisfy zero displacements at the pin A and zero vertical displacements at the pin on rollers at B, but setting the flexibilities associated with the forces at pin C to be zero only ensured that the rigid body displacements at C were consistent with the displacements at the other pins. Note that since the displacement at A is zero, the displacement at B is also the elongation, Δ_{AB}, of the spring AB so that the tension in that spring should be $T_2 = k\Delta_{AB} = ku_B = 48$ lb which is indeed the tension found previously.

If we find the displacements through the pseudo stiffness matrix, we need to modify the flexibility matrix to replace the zero flexibilities with small values:

```
flex = 1/k* [ 1 1 10^-6 10^-6 10^-6 10^-6 10^-6] ;
G = diag(flex) ;
Kp = E*inv(G)*E.';   % form stiffness
U = Kp\P;            % solve for displacements
U
 U =
 -38400121800081/(1600002880000*k)
 64000233000189/(2000003600000*k)
 12800040600027/(3200005760000*k)
-38400083400081/(1600002880000*k)
 64000143000027/(2000003600000*k)
-12800010599973/(3200005760000*k)
```

These approximate symbolic results are hard to read, but to a very good approximation they are:

```
U =
-24/k
32/k
4/k
-24/k
32/k
-4/k
```

which agrees with our previous solution.

In obtaining this solution, there were many non-deformable forces involved. We also can solve this problem for the tensions in the springs only, which makes the problem very direct, so let's consider that case.

(continued)

Example 2.4 (continued)

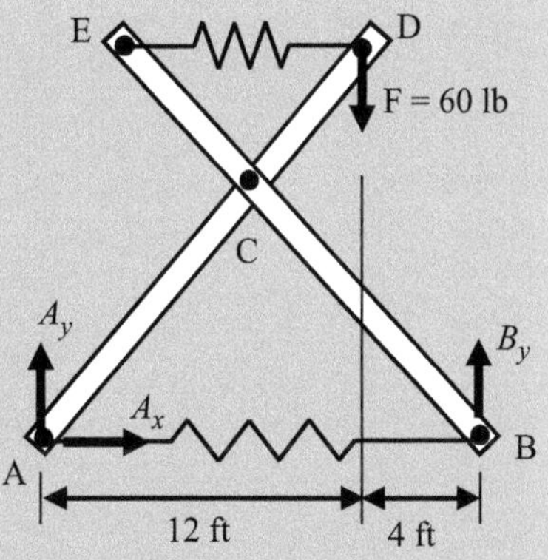

Fig. 2.12 A free body diagram of the entire X-frame

If we examine a free body diagram of the entire structure (Fig. 2.12), we see that we can solve for all the reaction forces. For the entire structure, summing forces and taking moments about point B we have:

$$\begin{aligned} &\sum F_x = 0 \quad A_x = 0 \\ &\sum F_y = 0 \quad A_y + B_y - 60 = 0 \\ &\sum M_{zB} = 0 \quad -16A_y + (4)(60) = 0 \\ &\rightarrow A_y = 15 \text{ lb}, \quad B_y = 45 \text{ lb} \end{aligned} \tag{2.51}$$

If we next examine the free body diagram of member ACD (Fig. 2.11a), we see that there are now four unknowns, since A_x and A_y have been obtained, but only three equations of equilibrium, and so the problem is statically indeterminate of order one. But if we only take moments about C for ACD, we obtain one equation for only the two unknown spring forces. This also requires one additional compatibility equation for a solution. This equilibrium equation is:

$$\begin{aligned} &\sum M_{zC} = 0 \quad 6T_2 + 3T_1 - (4)(60) - (8)(15) = 0 \\ &\rightarrow \quad 3T_1 + 6T_2 = 360 \end{aligned} \tag{2.52}$$

Our solution in MATLAB now is very simple. We can form up the equilibrium equation directly in matrix-vector form and perform the rest of the solution:

```
syms k
E = [ 3 6];    % equilibrium matrix
P = 360;   % force vector
S = null(E).';   % compatibility matrix
G = [1/k 0; 0 1/k];   % flexibility matrix (no zeros)
```

(continued)

Example 2.4 (continued)

```
Esys = [ E; S*G];   % set up system equations
Psys = [P;0];
F = Esys\Psys       % solve for T1, T2
F = 24
    48
```

we find again $T_1 = 24$ lb, $T_2 = 48$ lb, which are independent of k. If we want to solve for the one displacement here, we have:

```
U = Esys.'\G*F
 U = 8/k
     0
```

which gives $\{U\} = 8/k$ and a zero. What, physically, is the displacement here? We see from Fig. 2.13 that regardless of the angle of the X-frame, we have the lengths of the springs given by $L_1 = 10 \sin \theta$ and $L_2 = 20 \sin \theta$, so that for small changes of those lengths, we also have the deformations (elongations) of the springs given by $\Delta_1 = \Delta L_1 = 10 \cos \theta \Delta\theta = 6\Delta\theta$ and $\Delta_2 = 20 \cos \theta \Delta\theta = 12\Delta\theta$ for the configuration of Fig. 2.10 where $\cos\theta = 3/5$. Comparing this with the strain-displacement relation we find:

$$\begin{Bmatrix} \Delta_1 \\ \Delta_2 \end{Bmatrix} = \begin{bmatrix} 3 \\ 6 \end{bmatrix} \{U\} = \begin{bmatrix} 3 \\ 6 \end{bmatrix} (2\Delta\theta) \tag{2.53}$$

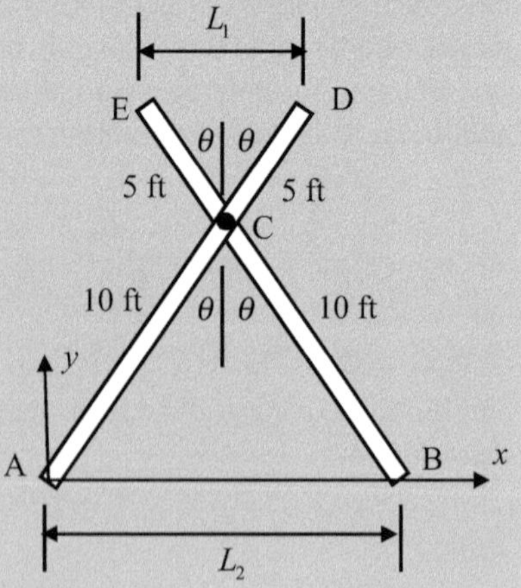

Fig. 2.13 The X-frame of Fig. 2.10 in a general configuration

(continued)

Example 2.4 (continued)
so $\{U\} = 2\Delta\theta$, i.e., it is the clockwise change of the total angle $\Delta\alpha = 2\Delta\theta$ between the two arms of the X-frame (measured in radians). This is a rotation about the negative z-axis, i.e., $\Delta\alpha = 2\Delta\theta = -2\Delta\theta_z$, which shows a negative sign again, as explained earlier. This angle involves both members ACD and BCE, so it shows how the physical meanings of the displacements change, depending on the equilibrium equations used, as mentioned earlier.

References

1. Schmerr, L. W. (2021). *Advanced mechanics of solids*. Cambridge University Press.
2. Li, G. (2020). *Introduction to the finite element method and implementation with MATLAB*. Cambridge University Press.
3. Kaljević, I., Patnaik, S. N., & Hopkins, D. A. (1996). *Development of finite elements for two-dimensional structural analysis using the integrated force method*. NASA Technical Memorandum 4655.
4. Kaljević, I., Patnaik, S. N., & Hopkins, D. A. (1996). *Element library for three-dimensional stress analysis by the integrated force method*. NASA Technical Memorandum 4686.

Chapter 3
Force-Based Methods in Statics

Objectives

- To introduce a second-generation force-based method at the level of a statics course.
- To give examples of the solution of statically indeterminate problems like the statically determinate problems normally considered in a statics course.
- To demonstrate the use of MATLAB in obtaining second-generation force-based solutions without programming where MATLAB is used as an engineering "calculator."

3.1 Introduction

Engineering statics is one of the oldest engineering subjects. It is also a foundational course in many engineering curricula. The primary focus of statics courses is to introduce the concept of equilibrium and solve equilibrium equations for the forces/moments acting on rigid bodies. The problems considered are statically determinate, i.e., the number of unknown forces/moments is equal to the number of equilibrium equations. Statics courses have traditionally relied on a background in calculus, trigonometry, and some aspects of linear and vector algebra. To include a discussion of statically indeterminate problems with force-based methods in statics, it is useful to also introduce matrices and matrix algebra, but the additional background needed in those topics is modest and can be easily incorporated into a statics course. Such a background can also be very valuable in supporting later courses since matrices and matrix algebra have become integral parts of many engineering subjects. By also including symbolic algebra in the force-based solution procedure, solving both statically determinate and statically indeterminate problems, as shown in the previous chapter, is made very simple and focuses a student's attention to drawing free

L. W. Schmerr Jr., *Algebraic Equations of Linear Elasticity*,
https://doi.org/10.1007/978-3-031-66174-7_3

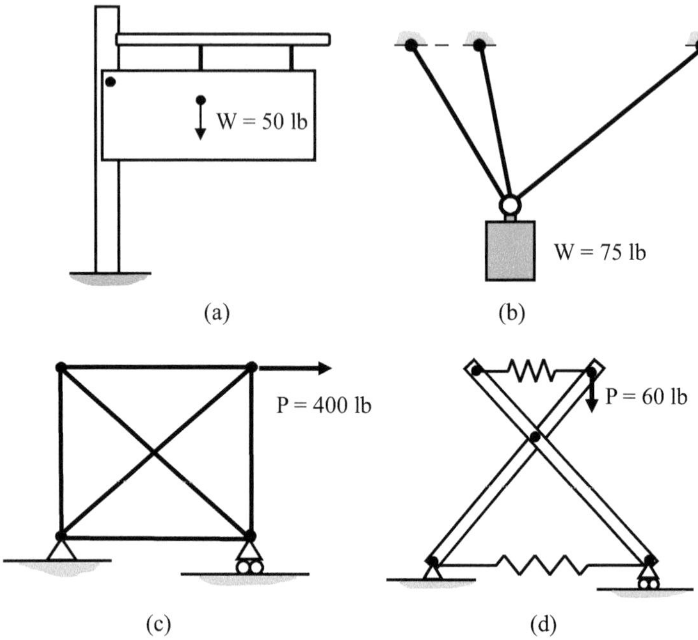

Fig. 3.1 Statically indeterminate problems. (**a**) Hanging sign. (**b**) A wire-supported weight. (**c**) A truss. (**d**) A pin-connected frame with springs

body diagrams and establishing equilibrium equations from those diagrams, which are skills identified as essential to a statics course. Combined with the ability of MATLAB to seamlessly perform matrix algebra, statics problems can be solved without any programming, effectively using MATLAB as a modern and powerful engineering "calculator."

Traditionally, statics has solved only statically determinate rigid body problems, but there are similar problems that are statically indeterminate that can also be treated at the technical level of a statics course. These include systems supported by linear elastic wires (Fig. 3.1a, b), trusses where the members are modeled as linear elastic two-force members (Fig. 3.1c), and pin-connected rigid members or frames containing linear elastic spring elements like the configuration we examined in Chap. 2 (Fig. 3.1d). In these problems, the deformable members are assumed to be very stiff so that the deformations present can be ignored when drawing the free body diagrams, as demonstrated in the last chapter. This assumption is satisfied for many structures and many structures also often behave in a linear elastic manner for small deformations. All the examples in Fig. 3.1 are two-dimensional (2D) geometries, but there are corresponding three-dimensional problems that can be treated in the same manner. Now, let us describe the general second-generation force-based approach specifically for some 2D problems.

3.2 Elements of a Force-Based Approach for Statics Problems

In obtaining the algebraic equations of linear elasticity recall, we used the fact that the stresses in the body are represented in terms of discrete "force" parameters, $\{F\}$, and must satisfy homogeneous equilibrium equations. Any distributed body forces present, such gravity acting on the body, are transformed into equivalent discrete applied forces $\{Q\}$, as are any distributed applied tractions (see Eq. (2.16)). We will say more about this transformation later. Elastic wires, elastic truss members, and springs are all modeled as weightless two-force members, where there are only discrete forces acting on the ends of the member, so we do not have to consider replacements of any distributed forces. In these cases, the internal force, F, is a constant axial force in the entire member. If, however, we also had axial distributed forces $q(x)$ (having the dimensions of force/unit length) acting on the member (Fig. 3.2a), then from equilibrium of a small element (Fig. 3.2b), we find

$$\begin{aligned} &\sum F_x = 0 \left(F(x) + \frac{dF}{dx} dx \right) - F(x) + q(x)dx = 0 \\ &\rightarrow \frac{dF}{dx} = -q(x) \end{aligned} \tag{3.1}$$

which is a differential equation of equilibrium for the internal force, F. The homogeneous solution where $q = 0$ is just $F = F_0 =$ constant (see Fig. 3.2c), so the constant internal forces present in all the elastic members shown in Fig. 3.1 are indeed solutions of homogeneous equilibrium equations for those members and the applied forces and reactions acting on the deformable elements are discrete forces acting at the ends (nodes) of the members so that they satisfy two of the four

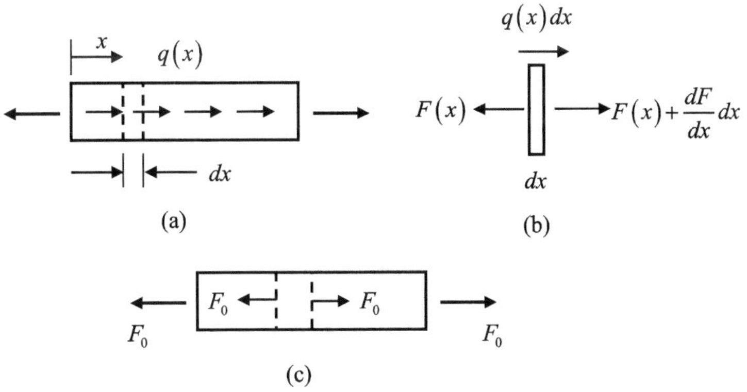

Fig. 3.2 (**a**) A bar in equilibrium under the action of end forces and a distributed force/unit length along the bar. (**b**) A free body diagram of a small element of the bar, where $F(x)$ is the internal axial force acting in the bar. (**c**) A bar in equilibrium under the action of only end forces, where the internal force is a constant

assumptions listed in Chap. 2 that led to the development of the algebraic equations of elasticity. We will again discuss axial force problems in the next chapter where distributed forces acting on the elastic members are present, and we will also solve those problems with force-based methods. In engineering statics courses, however, it is useful to keep the discussion as simple as possible and only consider problems like those shown in Fig. 3.1.

In the force-based approach we need to find the governing equations of equilibrium in the matrix-vector form $[E]\{F\} = \{P\}$. In statics problems, the unknown forces $\{F\}$ may include forces in deformable members as well as reactions from fixed (rigid) supports or connections. We saw in the last chapter how to include such rigid supports. For the structures shown in Fig. 3.1, these equations of equilibrium are typically obtained by examining the forces acting at the ends or common joints (pins) of the members.

In addition to the equilibrium equations, one also needs the flexibility matrix, $[G]$, that defines the relationship between the deformations and the forces present in a structure. For a linear elastic element such as a wire or a truss member the axial deformation (elongation), Δ, of the element is proportional to the internal axial force, F, where

$$\Delta = \frac{L}{AE} F \tag{3.2}$$

and where L is the length of the member, A is its cross-sectional area, and E is Young's modulus (a material property) having the dimensions of force/unit area. For steel, for example, $E = 30 \times 10^6$ lb./in^2 or, equivalently, $E = 210 \times 10^9$ N/m^2. These large material property values typically mean that the deformations are extremely small, so when obtaining the equilibrium equations from free body diagrams we can treat all the structural elements as if they were rigid. This was another assumption listed in Chap. 2. The quantity $g = \frac{L}{AE}$ is called *flexibility* (usually having a dimension of in/lb. or mm/N, for example), and the inverse of this quantity, $k = 1/g = AE/L$, is called the *stiffness* (having a dimension of lb./in or N/mm, for example). For linear elastic springs like those discussed in the previous chapter, the force-deformation is often written in terms of the stiffness (also called the *elastic spring constant*) as $F = k\Delta$ rather than the flexibility $g = 1/k$. As seen in the algebraic equations of linear elasticity, we need to incorporate these member flexibilities into a flexibility matrix for all the forces $\{F\}$ in the equations of equilibrium and their corresponding deformations $\{\Delta\}$. For a system where all the linear elastic elements are two-force members or springs, the flexibility matrix is just a diagonal matrix, $[G]$, containing the flexibilities associated with all the internal elastic forces acting on the main diagonal. If there are other forces in the equilibrium equations not associated with deformations, we saw in the last chapter that one needs to set their flexibilities equal to zero to ensure they represent forces at rigid supports or connections. For example, we might have

$$\{\Delta\} = [G]\{F\} \quad \rightarrow \quad \begin{Bmatrix} \Delta_1 \\ \Delta_2 \\ \Delta_3 \end{Bmatrix} = \begin{bmatrix} g_1 & 0 & 0 \\ 0 & g_2 & 0 \\ 0 & 0 & 0 \end{bmatrix} \begin{Bmatrix} F_1 \\ F_2 \\ R \end{Bmatrix} \tag{3.3}$$

if the forces F_1, F_2 were internal elastic forces and R was a reaction force at a rigidly fixed support. Setting the flexibility equal to zero for a force at a fixed support also guarantees there is no displacement associated with that force, as demonstrated in the last chapter.

The equations of equilibrium and the flexibility matrix are all that we need to implement the force-based method for a statically indeterminate problem since the compatibility matrix, [*S*], can be obtained from solutions of the homogeneous equilibrium equations, allowing us to form up a solvable system of equations (see Eq. (2.33)):

$$\begin{bmatrix} [E] \\ - \\ [S][G] \end{bmatrix} \{F\} = \begin{Bmatrix} \{P\} \\ - \\ \{0\} \end{Bmatrix} \tag{3.4}$$

The flexibility matrix, as just shown, is easy to obtain (and understand), so if one sticks to problems such as those shown in Fig. 3.1, including them in a statics course is appropriate to the level of the course. More complex flexibilities are best left to later courses. Equilibrium equations are already a concept familiar to students in statics, so in a force-based approach the only thing new in that part of the solution is placing those equations in the matrix-vector form $[E]\{F\} = \{P\}$. However, we saw in the last chapter that by using the symbolic algebra capabilities of MATLAB both $[E]$ and $\{P\}$ can be seamlessly obtained from the equilibrium equations expressed symbolically in the same form as when written down on paper. Obtaining the compatibility matrix [*S*] from homogeneous solutions of the equilibrium equations is a more abstract concept for statics students, so for simple problems, it is best to first obtain [*S*] by drawing diagrams of the deformations and corresponding displacements and showing how the compatibility equations arise from those diagrams. Such diagrams can also justify why the deformations and displacements are related through the transpose of the equilibrium matrix, i.e., $\{\Delta\} = [E]^T\{U\}$, which is also not an obvious relationship for students. As seen in Chap. 1, it is this relationship that leads to $[E][S]^T = [0]$, where the columns of [S], as discussed in that chapter, are the homogeneous solutions of the equilibrium equations.

3.3 Example Statics Problems

First, examine a problem that is ideal for implementing a force-based method. Because the problem is simple, we do not need to use symbolic algebra, so we will first give a solution with numerical matrix-vector algebra only.

Example 3.1: Wire-Supported Beam

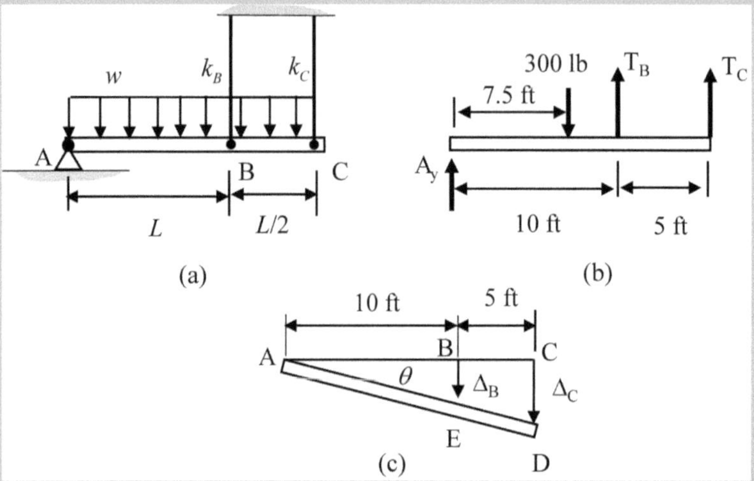

Fig. 3.3 (**a**) A rigid bar supported by elastic wires. (**b**) A free body diagram of the bar for $L = 10$ ft. (**c**) A diagram of the rotational displacement of the bar and the elongations of the wires

Consider the wire-supported beam shown in Fig. 3.3a. The distance $L = 10$ ft. The beam carries a uniform load $w = 20$ lb/ft and the wires deform in a linear elastic manner. A free body diagram of the beam is shown in Fig. 3.2b where the distributed force has been replaced by its resultant. The stiffnesses of the wires are $k_B = 1200$ lb/in and $k_C = 1000$ lb/in. The beam is assumed to be rigid while the elastic wires deform. The change of geometry is shown in Fig. 3.3c.

From the free body diagram of Fig. 3.3b the equilibrium equations are

$$\begin{aligned} \sum F_y &= 0 \quad A_y + T_B + T_C - 300 = 0 \\ \sum M_{zA} &= 0 \quad 10T_B + 15T_C - (7.5)(300) = 0 \end{aligned} \tag{3.5}$$

where we have taken moments about point A. The beam is shown as horizontal on the free body diagram because any deformations of the wires are assumed to be very small. The force equation involves forces associated with the elastic wires as well as the reaction force at the rigid support while the moment equation only has forces associated with the deformable elements. Let us first use only the moment equation. This is one equation for the two unknown tensions in the wires so it is statically indeterminate. In matrix-vector form we have

$$[E]\{F\} = \{P\} \rightarrow [10 \;\; 15]\begin{Bmatrix} T_B \\ T_C \end{Bmatrix} = \{2250\} \tag{3.6}$$

(continued)

Example 3.1 (continued)

Now, consider the deformations (Δ_B, Δ_C) of the wires shown in Fig. 3.3c and the angular displacement (rotation), θ, of the beam. From the geometry we see that for a small rotation, the deformations are given by $\Delta_B = 10\theta$, $\Delta_C = 15\theta$, so that in matrix-vector form, the deformations and displacement are related through the transpose of the equilibrium matrix, i.e.,

$$\{\Delta\} = [E]^T\{U\} \quad \rightarrow \quad \begin{Bmatrix} \Delta_B \\ \Delta_C \end{Bmatrix} = \begin{bmatrix} 10 \\ 15 \end{bmatrix}\{\theta\} \tag{3.7}$$

Thus, in the deformation-displacement relation, the transpose of the equilibrium matrix does appear. Also, from the geometry changes, we see from the similar triangles ABE and ACD, the deformations are related through

$$\frac{\Delta_B}{10} = \frac{\Delta_C}{15} \tag{3.8}$$

which yields the compatibility equation:

$$[S]\{\Delta\} = \{0\} \quad \rightarrow \quad [-3/2 \quad 1]\begin{Bmatrix} \Delta_B \\ \Delta_C \end{Bmatrix} = 0 \tag{3.9}$$

This is a very direct and visual approach, but it is not practical for more complex problems and is not amenable to being automated on computers. However, in a statics course it gives students a valuable method to obtain compatibility equations and understand their significance. An approach that is viable for more complex problems is to obtain the compatibility matrix $[S]$ from the homogeneous solution of this equilibrium equation, as discussed in Chap. 1. In statics, students do not have to solve a homogeneous system of equations by hand because we can do this with the MATLAB null function. When we take the matrix transpose of those homogeneous solutions we obtain $[S]$:

```
E = [10 15];
S = null(E, 'r').'
S = -1.5000 1.0000
```

This result agrees with Eq. (3.9). One can use other simple sets of homogeneous equations to demonstrate how such solutions can be found [1]. Here we used the rational option 'r' in the null function so that the MATLAB result would be in whole numbers or fractions, as found in Eq. (3.9). Generally,

(continued)

Example 3.1 (continued)
when solving problems, we will omit this option. As mentioned in Chap. 2, we always have some freedom in writing the homogeneous compatibility equations since we can multiply them by arbitrary constants or take different combinations (if there are more than one compatibility equations) and still have a legitimate result.

Here is the complete solution in MATLAB:

```
E = [10 15];
P = 2250;
S = null(E).';
G = [ 1/1200 0; 0 1/1000];
Esys = [E; S*G];
Psys = [P;0];
F = Esys\Psys
F = 78.2609               % tension Tb (lb)
    97.8261               % tension Tc (lb)
Ay = 300 - T(1) - T(2)    % reaction at A (lb) from force equilibrium
Ay = 123.9130
```

This solution used just the moment equation and solved for the tensions. Force equilibrium was then used to obtain the reaction at A. The null function was used without the 'r' argument. Note that the units in the moment equation were foot-pounds while the stiffnesses were measured in pounds/inch. This inconsistency is acceptable since, as mentioned previously, we can always change the units of the terms in the flexibility matrix by multiplying by a constant without affecting the compatibility equations. When obtaining displacements, however, one must use consistent units in the equilibrium equation(s) and the compatibility equation(s), as we have discussed previously.

One could also solve this problem for all the forces using both the force and moment equilibrium equations, (Eq. 3.5), which we rewrite here:

$$\begin{aligned} \sum F_y = 0 \quad & A_y + T_B + T_C - 300 = 0 \\ \sum M_{zA} = 0 \quad & 10T_B + 15T_C - (7.5)(300) = 0 \end{aligned} \tag{3.10}$$

so we have

$$[E]\{F\} = \{P\} \quad \rightarrow \quad \begin{bmatrix} 1 & 1 & 1 \\ 0 & 10 & 15 \end{bmatrix} \begin{Bmatrix} A_y \\ T_B \\ T_C \end{Bmatrix} = \begin{Bmatrix} 300 \\ 2250 \end{Bmatrix} \tag{3.11}$$

(continued)

Example 3.1 (continued)

In this case, we need to form a flexibility matrix with zero flexibility for the reaction force at A. This is easy to understand since there is no deformation associated with that support. Setting the flexibility to zero at A will also ensure the displacement is zero at that point, as demonstrated in Chap. 2. Here is that solution, where now we will use symbolic algebra:

```
syms Ay Tb Tc
Eq(1) = Ay + Tb + Tc -300 == 0;
Eq(2) = 10*Tb+ 15*Tc -2250 == 0;
T= [ Ay Tb Tc];
[E, P] = equationsToMatrix(Eq, T);
S = null(E).';
G= [ 0 0 0;0 1/1200 0; 0 0 1/1000]; % note zero for flexibility
    associated with Ay
Esys = [E;S*G];
Psys = [P;0];
F = Esys\Psys;
Fn = double(F)
Fn = 123.9130 % reaction Ay (lb)
     78.2609 % tension Tb (lb)
     97.8261 % tension Tc (lb)
```

This obviously gives the same solution. In this case we wrote the equilibrium equations in symbolic form and then obtained the equilibrium matrix [E] and the force term {P}from the MATLAB function equationsToMatrix. This was not strictly necessary since we only have two equations here, but for problems with many more equations, this is a very effective approach.

As shown in Chap. 2, once the forces are known, we can also determine the displacements, or we can form a pseudo stiffness matrix and determine the displacements directly. In discussions of the force-based method in statics, one can limit the discussion of displacements to justifying the deformation-displacement relationship $\{\Delta\} = [E]^T\{U\}$ since finding forces/moments are the main emphasis in a statics course. We will, however, examine displacements for this problem later in this chapter.

Trusses are an important part of a statics course since they describe familiar structures like those found in many bridges. Statically determinate trusses can be solved by the method of pins (joints) with the equilibrium equations. The force-based method can also solve statically indeterminate trusses with little additional effort. Here is an example.

Example 3.2: Statically Indeterminate Truss

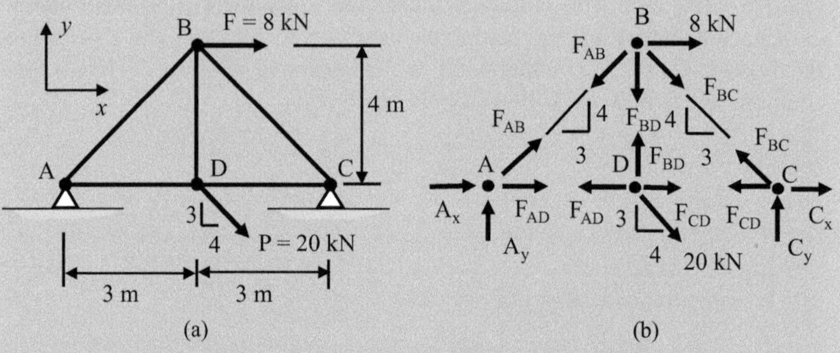

Fig. 3.4 (**a**) A statically indeterminate truss. (**b**) Free body diagrams of the pins

The statically indeterminate truss in Fig. 3.4a is fixed at both A and C and is loaded by the forces $F = 8$ kN and $P = 20$ kN. The flexibility of a truss member is $g = L/EA$ where A is the cross-sectional area, E is Young's modulus, and L is the member length. Assume EA is the same for all members. Determine the forces in the truss members and the reactions at points A and C. We will use the method of pins (joints) for this problem to obtain the equilibrium equations. We first need to draw the free body diagrams of all the pins, as shown in Fig. 3.4b. The eight equilibrium equations at pins A, B, C, and D are:

$$
\begin{aligned}
{}_A\sum F_x &= 0 \quad A_x + F_{AD} + 3F_{AB}/5 = 0 \\
{}_A\sum F_y &= 0 \quad A_y + 4F_{AB}/5 = 0 \\
{}_B\sum F_x &= 0 \quad -3F_{AB}/5 + 3F_{Bc}/5 + 8 = 0 \\
{}_B\sum F_y &= 0 \quad -F_{BD} - 4F_{AB}/5 - 4F_{BC}/5 = 0 \\
{}_C\sum F_x &= 0 \quad C_x - F_{CD} - 3F_{BC}/5 = 0 \\
{}_C\sum F_y &= 0 \quad C_y + 4F_{BC}/5 = 0 \\
{}_D\sum F_x &= 0 \quad F_{CD} - F_{AD} + 16 = 0 \\
{}_D\sum F_y &= 0 \quad F_{BD} - 12 = 0
\end{aligned}
\tag{3.12}
$$

There are nine unknowns, consisting of the five forces in the truss members and the four reaction forces at A and C. Thus, the problem is statically indeterminate. The compatibility matrix in this case is a vector. Note that we

(continued)

Example 3.2 (continued)
could solve the last equation for F_{BD} and hence eliminate that equation from consideration, but the problem remains statically indeterminate. We will keep all equations in our solution.

To solve this problem, we will use the symbolic algebra capabilities of MATLAB which allows us to write the equilibrium equations of Eq. (3.12) directly in symbolic form:

```
% symbolic member forces and reactions
syms Fab Fad Fbc Fbd Fcd Ax Ay Cx Cy
% force vector of all the unknowns
T = [Fab Fad Fbc Fbd Fcd Ax Ay Cx Cy];
% equations of equilibrium at the pins
Eq(1) = Ax + Fad +3*Fab/5 == 0;
Eq(2) = Ay +4*Fab/5 == 0;
Eq(3) = -3*Fab/5 + 3*Fbc/5 + 8 == 0;
Eq(4) = -Fbd -4*Fab/5 - 4*Fbc/5 == 0;
Eq(5) = Cx - Fcd -3*Fbc/5 == 0;
Eq(6) = Cy + 4*Fbc/5 == 0;
Eq(7) = Fcd -Fad +16 == 0;
Eq(8) = Fbd -12 == 0;
```

We can omit the symbolic equals zero terms, == 0, in these equations, but including them makes them correspond more closely with the written equations of Eq. (3.12). Once we have these symbolic equations, we can use them to define the equilibrium equations in matrix form [E]{T} = {P}. The MATLAB function equationsToMatrix gives us the equilibrium matrix [E] and the vector {P}:

```
[E, P] = equationsToMatrix(Eq, T);
```

We need the equilibrium matrix to determine the compatibility matrix. Having [E] we then can use the MATLAB null function, which obtains the solutions $\{T_h\}$ of the homogeneous equilibrium equations, $[E]\{T_h\} = \{0\}$. We output them in a matrix, $[S]^T$, that is the transpose of the compatibility matrix $[S]$. Taking the transpose of the matrix output of null then gives us the compatibility matrix $[S]$ itself:

```
S = null(E).';
```

(continued)

Example 3.2 (continued)

Next, we need to define the flexibility matrix [G]. We can omit a common flexibility term L_r/EA for all the truss members, where we will let a reference length be L_r = Lab = 5 m since such a common factor can always be eliminated from the homogeneous compatibility equations. Then the remaining terms in the flexibility matrix for the members are just ratios of the member lengths to L_r. We set the flexibilities of the reactions at A and C equal to zero since there are no deformations associated with these forces at the fixed pins.

```
% form up a vector of normalized flexibilities (length ratios) and
zeros
flex = [ 1 3/5 1 4/5 3/5 0 0 0 0];
% and place them on the diagonal of [G] with the MATLAB diag function
G = diag(flex);
```

We then combine the equilibrium and compatibility equations and form the elements of the system equations [Esys]{F} = {Psys}. We solve for the unknown forces, where the components in {F} are the values of the symbolic tensions contained in {T} vector defined previously:

```
Esys = [ E; S*G];
Psys = [P; 0];
F = Esys\Psys;
```

The vector F is still in symbolic form, so we convert it to a double precision numerical vector with the MATLAB function double:

```
Fn = double(F)
Fn =
  -0.8333    % Fab = - 0.8333 kN
   8.0000    % Fad = 8 kN
 -14.1667    % Fbc = -14.1667 kN
  12.0000    % Fbd = 12 kN
  -8.0000    % Fcd = -8 kN
  -7.5000    % Ax = -7.5 kN
   0.6667    % Ay = 0.6667 kN
 -16.5000    % Cx = -16.5 kN
  11.3333    % Cy = 11.3333 kN
```

(continued)

Example 3.2 (continued)

You can verify that the reactions do satisfy equilibrium for the entire truss. Note the directness of this solution. You only need to (1) write down the equilibrium equations in a form essentially identical to how they are written on paper and (2) define a vector of flexibilities. The remaining steps then all use the built-in symbolic algebra and numerical capabilities of MATLAB to obtain the solution.

3.4 Calculating Displacements

As seen in the two examples just discussed, one can solve problems for the forces where the compatibility equation is obtained either from a graphical diagram of the deformations and displacements or from homogeneous solutions of the equilibrium equations. In a statics course, where forces and moments play a central role, one could omit any detailed discussion of solving for the displacements, leaving those topics to later courses such as strength of materials. The materials in a forthcoming statics text and in its supporting web materials [1, 2], however, do include some discussions of how to solve for displacements. One method, as seen in the previous chapters, is to form up the equilibrium equations in terms of the displacements by combining the force-deformation and deformation-displacement relations and the equilibrium equations in the algebraic equations of elasticity to yield the equilibrium equations for the displacements, as discussed previously, in the form

$$[E][G]^{-1}[E]^T\{u\} = \{P\} \tag{3.13}$$

where $[K_p] = [E][G]^{-1}[E]^T$ is the pseudo stiffness matrix obtained from the equilibrium matrix and the inverse of the flexibility matrix. This is also an alternative displacement-based approach to completely solving this problem for both the displacements and the forces. We will consider this method as well as the one discussed previously for obtaining the displacements (see, for example, Eq. (1.30)), which can be used once the forces are found with the force-based methods without using a stiffness matrix. For cases where the equilibrium equations only involve internal forces associated with deformable elements, like the moment equation considered in the first example, the inverse of the flexibility matrix is well defined so that the use of Eq. (3.13) is straightforward. Thus, we only need the equilibrium equations and the flexibility matrix to determine the displacements. In cases where the equilibrium equations involve forces associated with rigid elements or connections, the inverse of the flexibility matrix does not formally exist, but if one replaces any zero flexibilities in the flexibility matrix by small non-zero values (i.e., small values relative to the other non-zero flexibilities which themselves are usually small), then one can invert the flexibility matrix and obtain a pseudo stiffness matrix that will guarantee to a good approximation that the displacements associated with rigid elements or connections are zero (see the previous discussion in Example 2.3).

Example 3.3: Wire-Supported Beam—Displacements
Now, consider the wire-supported bar of Fig. 3.3a with the use of only the moment equation and the pseudo stiffness matrix:

```
E = [10*12 15*12];  % equilibrium matrix (in-lb)
P = 2250*12;  % moment (in-lb)
G = [1/1200 0; 0 1/1000];  % flexibility matrix (in/lb)
K = E*inv(G)*E.';
u = K\P  % u = the angular displacement, theta
u = 5.4348e-04
F = inv(G)*E.'*u   % Tb and Tc in lb
F = 78.2609
    97.8261
```

We used the same dimensions of inches and pounds in both the moment equilibrium equation and the flexibility matrix to form up the stiffness matrix since this consistency is necessary when solving for the displacements. The single angular displacement obtained in this case was *the angle of rotation of the beam,* θ, but notice that we took moments about the positive z-axis to write the moment equation, so the angular displacement that should be obtained is $-\theta_z$, consistent with our previous discussion of the signs in Chap. 2 on the displacements obtained, and this is indeed the case since $\theta = -\theta_z$, where θ_z is the rotation of the bar about the positive z-axis. Once this displacement is known, we used the deformation-displacement relation, written as $\{F\} = [G]^{-1}[E]^T\{U\}$, to solve for the forces, which are the same forces found previously with the force-based approach.

If we solve the problem of Fig. 3.3a with both the force and moment equations instead, we have:

```
syms Ay Tb Tc
Eq(1) = Ay + Tb + Tc -300 ==0;
Eq(2) = 10*12*Tb + 15*12*Tc - 2250*12 == 0;
T = [ Ay Tb Tc];
[E, P] = equationsToMatrix(Eq, T);
G = [ 10^-9 0 0; 0 1/1200 0; 0 0 1/1000]; % modified flexibility matrix
K = E*inv(G)*E.';
u = K\P;
double(u)
ans = 1.0e-03 *
   0.0001          % y- displacement (approximately zero)
   0.5435          % displacement theta
F = inv(G)*E.'*u;  % forces Ay Tb Tc (lb)
```

(continued)

Example 3.3 (continued)

```
double(F)
ans = 123.9130
       78.2609
       97.8261
```

We see here that we modified the flexibility matrix with a very small value for the flexibility associated with Ay so that the inverse of the flexibility matrix could be obtained. The displacements obtained (approximately) were ($-u_{Ay}$, $-\theta_z$), where u_{Ay} is the y-displacement at A (which should be identically zero) and the $\theta = -\theta_z$ value found previously. The forces found from these displacements are approximately in agreement with those found previously, although the differences cannot be seen in the number of decimal places shown.

Using the stiffness matrix means that we are solving the equilibrium equations directly for the displacements. But if we have already found the forces with a force-based approach, as discussed previously, we can use those forces to find the displacements (see Eq. (2.34)), which we repeat here):

$$[E_{sys}]^T \left\{ \begin{array}{c} \{U\} \\ - \\ \{0\} \end{array} \right\} = [G]\{F\} \tag{3.14}$$

Once a force-based solution is obtained we already have $[G]$, $[E]$, $\{F\}$ and $[E_{sys}]$, so Eq. (3.14) can be solved for the displacements $\{U\}$ and one or more zeros in $\{0\}$. Note that the additional zeros do not represent any displacements present, which are all contained in the vector $\{U\}$. Since we used both $[G]$ and $[E]$ in this solution, recall it is essential that the units in these matrices be consistent. Also recall that $\{U\}$ will contain the negative of the displacements or angular displacements, as shown previously. Here is the solution for both the forces and displacements for the problem of Fig. 3.3a, using only the moment equation:

```
E = [10*12 15*12]; % [E] and {P} in terms of inches and pounds
P = 2250*12;
S = null(E).';
G = [1/1200 0; 0 1/1000];
Esys = [E; S*G];
Psys = [P;0];
F = Esys\Psys;
Fn = double(F)   % Tb and Tc (in lb)
```

(continued)

Example 3.3 (continued)

```
Fn = 78.2609
    97.8261
Ay = 300 - Fn(1) - Fn(2) % Ay (in lb)
Ay = 123.9130
u = Esys.'\G*F; % solution for {u}, which is theta (in radians), and
   a zero
double(u)
ans = 1.0e-03 *
     0.5435
     0.0000
```

In this case, we wrote the moment equation in inches and pounds to be consistent with the flexibility matrix. Similarly, if we use both the force and moment equations we have

```
syms Ay Tb Tc
Eq(1) = Ay + Tb + Tc -300 == 0;
Eq(2) = 10*12*Tb+ 15*12*Tc -2250*12 == 0;
T = [ Ay Tb Tc];
[E, P] = equationsToMatrix(Eq, T);
S = null(E).';
G = [ 0 0 0;0 1/1200 0; 0 0 1/1000];
Esys = [E;S*G];
Psys = [P;0];
F = Esys\Psys;
Fn = double(F)   % The force Ay, Tb, Tc (in lb)
Fn = 123.9130
 78.2609
 97.8261
u = Esys.'\G*F;
double(u)  % The negative displacement at A (zero), theta, and a
zero
ans = 1.0e-03 *
     0
     0.5435
     0
```

We find the negative *y*-displacement at A (which is zero) and the angular displacement, $\theta = -\theta_{yA}$, as well as an additional zero. Note that unlike the case where we used the stiffness matrix defined from a modified flexibility matrix that did not contain any zeros, here we can set the flexibility at A to be identically zero and obtain a zero displacement at that point without approximation.

The forces in a statically indeterminate truss can be easily found without a direct reference to the displacements, but it is not difficult to also determine the displacements at the pins of a truss, as shown in the next example.

Example 3.4: Statically Indeterminate Truss—Displacements
Consider now the truss problem of Fig. 3.4a. To obtain displacements we need to specify the flexibility term L_r/EA. We will let Young's modulus $E = \text{Ym} = 210 \times 10^9$ N/m^2, the cross-sectional area $A = 0.2$ m^2, and the reference length $L_r = 5$ m. The forces are all in kN. The MATLAB solution for first the forces and then the displacements using Eq. (3.14) is:

```
% symbolic member forces and reactions
syms Fab Fad Fbc Fbd Fcd Ax Ay Cx Cy

% Ym = Young's modulus (N/m^2), area A (m^2), Lr (m)
Ym = 210*10^9; A = 0.2; Lr = 5;

% force vector of all the unknowns
T = [Fab Fad Fbc Fbd Fcd Ax Ay Cx Cy];

% equations of equilibrium at the pins
Eq(1) = Ax + Fad +3*Fab/5 == 0;
Eq(2) = Ay +4*Fab/5 == 0;
Eq(3) = -3*Fab/5 + 3*Fbc/5 + 8 == 0;
Eq(4) = -Fbd -4*Fab/5 - 4*Fbc/5 == 0;
Eq(5) = Cx - Fcd -3*Fbc/5 == 0;
Eq(6) = Cy + 4*Fbc/5 == 0;
Eq(7) = Fcd -Fad +16 == 0;
Eq(8) = Fbd -12 == 0;
[E, P] = equationsToMatrix(Eq, T);
S = null(E).';

% form up a vector of flexibilities and zeros
flex = (Lr/(Ym*A))*[ 1 3/5 1 4/5 3/5 0 0 0 0];
% and place them on the diagonal of [G] with the MATLAB diag function
G = diag(flex);
Esys = [ E; S*G];
Psys = [P; 0];
F = Esys\Psys;    % forces in kN
Fn = double(F)
```

(continued)

Example 3.4 (continued)

```
Fn =

  -0.8333
   8.0000
 -14.1667
  12.0000
  -8.0000
  -7.5000
   0.6667
 -16.5000
  11.3333
u = Esys.'\G*(10^3)*F; % Place forces in N to get displacements in m
un = -10^3*double(u) % convert displacements to mm and change signs
un =
         0   % ux at A (mm)
         0   % uy at A (mm)
    0.0013   % ux at B (mm)
   -0.0011   % uy at B (mm)
         0   % ux at C (mm)
         0   % uy at C (mm)
    0.0006   % ux at D (mm)
   -0.0023   % uy at D (mm)
         0   % extra zero
```

Here we let Young's modulus be the variable Ym so as not to conflict with the symbol E used for the equilibrium matrix. With the given units, the flexibility term L_i/EA is measured in meters/Newton. The forces were measured in kN, so in the term [G]{F}, the force F values were multiplied by 10^3 to convert those forces to Newtons. Since the dimensions of the flexibility matrix are in meters/Newton, the resulting displacements are then measured in meters. The final two conversions seen in the MATLAB solution are to (1) change the signs on the displacements, so they are all displacements in the positive x- and y-directions, and (2) multiply the displacements by 10^3 to convert them to mm. The equilibrium equations were force equations in the x- and y-directions at pins (A, B, C, D), in that order, so the final displacements obtained are the displacements in the x- and y-directions at (A, B, C, D) in that same order, as indicated above. The final displacement value obtained above is an additional zero "padding". The displacements are all small, as expected. The displacements at the pins in the truss are given in Fig. 3.5a and the (greatly exaggerated) deformed shape of the truss is shown in Fig. 3.5b.

(continued)

Example 3.4 (continued)

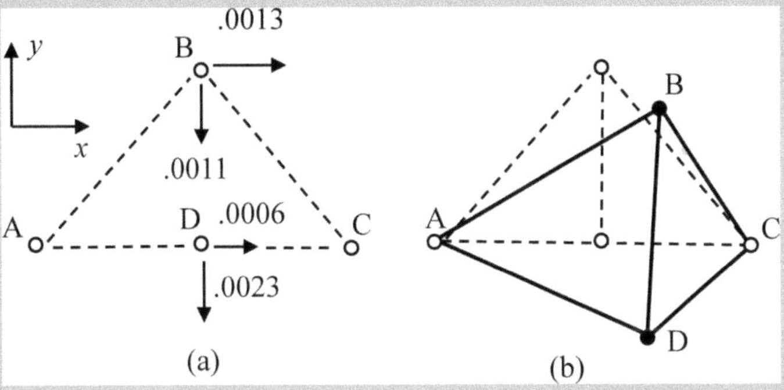

Fig. 3.5 (**a**) The displacements at the pins in their original (undeformed) positions. (**b**) The deformed shape of the truss (greatly exaggerated)

3.5 Spatial Trusses

Two-dimensional truss problems are excellent examples for applying force-based methods with MATLAB in a statics course. Spatial (3D) trusses can demonstrate the value of MATLAB to an even greater degree because of the use of vectors in setting up and solving the equilibrium equations. Here is an example.

Example 3.5: A Statically Indeterminate Spatial Truss
The four-bar spatial truss of Fig. 3.6 is supported by four ball and socket joints at A, B, D, and E and carries a 600 lb force in the negative z-direction and a 500 lb force in the y-direction. Members DC and EC lie in the x-y plane. Pin C is at the origin of the x-y-z coordinates and the coordinates of A are (−4, −1, 2) ft, the coordinates of B are (−4, 1, 2) ft., while the coordinates of D are (−4, −1, 0) ft and of E are (−4, 1, 0) ft. The flexibility of each truss member is given by

$$g = L/EA$$

where A is the cross-sectional area, E is Young's modulus, and L is the length. Let EA be the same for all members. Determine the forces in the truss members and the reactions at A, B, D, and E.

(continued)

Example 3.5 (continued)

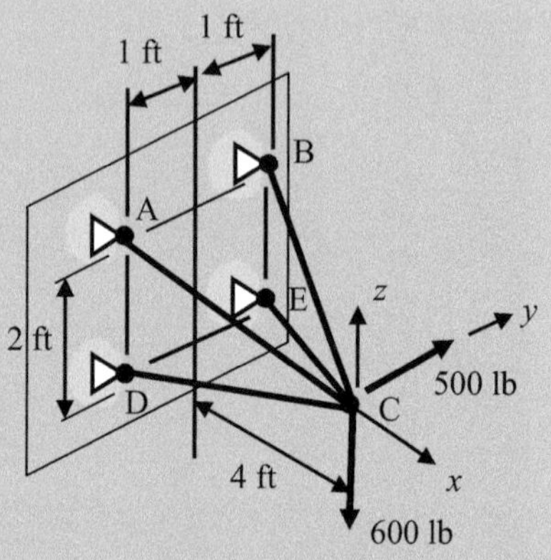

Fig. 3.6 A statically indeterminate spatial truss

Free body diagrams of the pins are shown in Fig. 3.7. There are 12 unknown reaction forces and 4 unknown forces in the members for a total of 16 unknowns. There are three equilibrium equations at each of the 5 pins, so there is a total of only 15 equilibrium equations and the truss is statically indeterminate. To determine the forces at the pins, we need to first find the unit vectors along the members. This is easily done in MATLAB:

```
CA = [-4 -1 2]; Lca = norm(CA); eCA = CA/norm(CA); eAC = -eCA;
CB = [-4 1 2]; Lcb = norm(CB); eCB = CB/norm(CB); eBC = - eCB;
CD = [ -4 -1 0]; Lcd = norm(CD); eCD = CD/norm(CD); eDC = - eCD;
CE = [ -4 1 0]; Lce = norm(CE); eCE = CE/norm(CE); eEC = - eCE;
```

Next, we need to define all the unknowns symbolically and place them in a vector T

```
syms Tca Tcb Tce Tcd Ax Ay Az Bx By Bz Dx Dy Dz Ex Ey Ez
T = [ Tca Tcb Tce Tcd Ax Ay Az Bx By Bz Dx Dy Dz Ex Ey Ez];
```

We will also define vectors of the reactions and the applied forces

```
A = [Ax Ay Az]; B = [Bx By Bz]; D = [Dx Dy Dz]; E = [Ex Ey Ez];
Pv = [0 500 -600];
```

We can form the equilibrium equations at each of the pins vectorially and place them all in the vector Eq:

(continued)

Example 3.5 (continued)

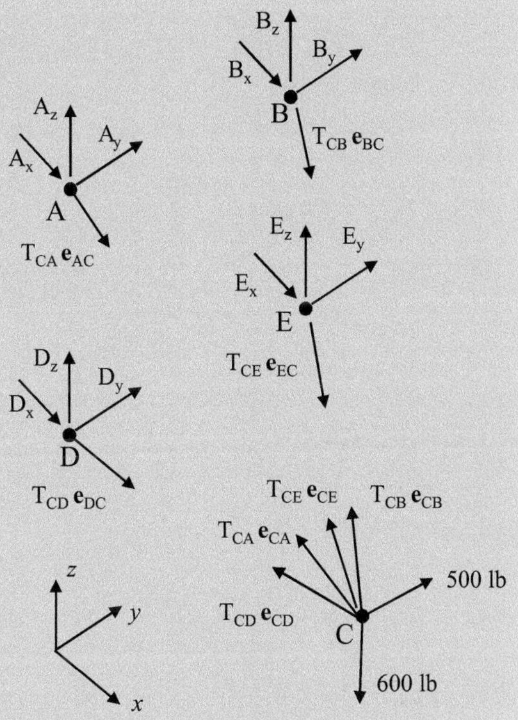

Fig. 3.7 Free body diagrams of the pins of the truss

```
EqC = Tca*eCA + Tcb*eCB + Tcd*eCD + Tce*eCE +Pv;
EqB = B + Tcb*eBC;
EqA = A + Tca*eAC;
EqD = D + Tcd*eDC;
EqE = E + Tce*eEC;
Eq = [ EqA EqB EqC EqD EqE];
```

These vector equations can then be converted to an equilibrium matrix [E] and a vector {P}, where [E]{T} = {P}. Note that we omitted the == 0 in writing the vector equations.

```
[E, P] = equationsToMatrix(Eq,T);
```

Next we find the compatibility matrix S:

```
S = null(E).';
```

(continued)

Example 3.5 (continued)

We factor out a common flexibility *a*/*EA* from the flexibility matrix, where a = 4 ft (an arbitrary choice), so the matrix will only contain ratios of the member lengths to a:

```
a = 4;
flex = [ Lca/a Lcb/a Lce/a Lcd/a 0 0 0 0 0 0 0 0 0 0 0 0];
G = diag(flex);
```

We now can combine the equilibrium and compatibility equations and solve for the forces, which are converted to numerical values

```
Esys = [ E; S*G];
Psys = [P; 0];
F = double(Esys\Psys)
F =
1.0e+03 *
   1.1702   % Tca = 1170.2 lb
   0.2046   % Tcb = 204.6 lb
  -1.2149   % Tce = -1214.9 lb
  -0.0221   % Tcd = - 22.1 lb
  -1.0214   % Ax = -1021.4 lb
  -0.2554   % Ay = -255.4 lb
   0.5107   % Az = 510.7 lb
  -0.1786   % Bx = -178.6 lb
   0.0446   % By = 44.6 lb
   0.0893   % Bz = 89.3 lb
   0.0214   % Dx = 21.4 lb
   0.0054   % Dy = 5.4 lb
   0        % Dz = 0 lb
   1.1786   % Ex = 1178.6 lb
  -0.2946   % Ey = -294.6 lb
   0        % Ez = 0 lb
```

You can easily check that force equilibrium is satisfied for the entire truss since

$$A_x + B_x + D_x + E_x = 0$$
$$A_y + B_y + D_y + E_y = -500$$
$$A_z + B_z + D_z + E_z = 600$$

Note the directness of this solution and its relative simplicity even though there were many unknowns. It uses the vector-matrix capabilities of MATLAB to their fullest extent.

A source of additional information on statically indeterminate statics problems and on the role of MATLAB in statics can be found in [2], which is a companion website in support of the new statics text [1].

References

1. Schmerr, L. W. (2024). *Engineering statics with MATLAB*. CRC Press.
2. www.eng-statics.org

Chapter 4
Force-Based Methods in Strength of Materials

Objectives

- To present the governing differential equations for the axial extension and torsion of bars.
- To define discrete forces/torques at nodes that are equivalent to distributed forces or distributed torques.
- To obtain force-based solutions to axial extension and torsion problems.
- To introduce singularity functions for plotting and evaluating forces/torques and displacements/rotations.
- To obtain displacement-based solutions with a pseudo stiffness matrix.
- To discuss the relationship between the force-based approach presented in this chapter and force-based finite elements.

4.1 Introduction

A strength of materials course typically follows a statics course in an engineering curriculum. Concepts of stress and strain and stress-strain relations are introduced in this course, and they are applied primarily to three types of deformable bodies: (1) axially loaded bars, (2) torsion of bars with a circular cross-section, and (3) the bending of beams under lateral loads. Figure 4.1a–c gives examples of each of these problem types where the problems shown are all statically indeterminate and have a mixture of distributed and concentrated applied loads.

We can solve all the problems shown in Fig. 4.1 with second-generation force-based methods. However, the discussion of beam problems is more complex, so they will be covered in Chap. 5 at an advanced strength of materials level. Chapter 5, however, will also show how all the elements of a force-based solution to beams can be included in a strength of materials course. Currently, statically indeterminate

L. W. Schmerr Jr., *Algebraic Equations of Linear Elasticity*,
https://doi.org/10.1007/978-3-031-66174-7_4

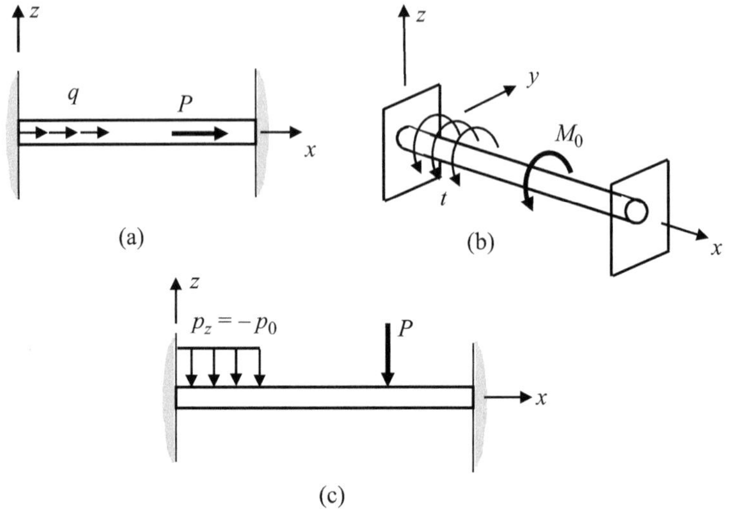

Fig. 4.1 (**a**) A bar with applied distributed and concentrated axial forces. (**b**) A bar with applied distributed and concentrated torques. (**c**) A beam with applied distributed and concentrated lateral forces

problems in strength of materials courses are typically solved with displacement-based methods. The axial extension and torsion problems considered in this chapter will introduce a new perspective on how to solve these problems directly for the forces, using second-generation force-based methods that are also applicable to all courses in a solid mechanics curriculum. The force-based method for strength of materials problems will also be used to illustrate some of the elements of a force-based finite element method, thus laying the background for a later finite elements course.

4.2 Axial Forces

4.2.1 Governing Differential Equations

When bars are loaded axially with the internal axial force (and applied forces) passing through the centroid, *C*, of the cross-section of a bar, as shown in Fig. 4.2, the only internal stress generated in the bar is the stress σ_{xx} and that stress is only a function of *x*, i.e., $\sigma_{xx} = \sigma_{xx}(x)$ so that the stress is uniform across the cross-section of the bar and the internal axial force is $F(x) = \sigma_{xx}(x)A$, where *A* is the cross-sectional area. The area *A* is assumed here to be a constant. The bar is shown with a rectangular cross-section in Fig. 4.2, but the bar could have other cross-sectional shapes. There is also an axial strain, $e_{xx}(x)$, acting in the bar and that axial strain is related to the displacement in the *x*-direction, $u(x)$, $e_{xx} = du/dx$. Physically, this strain

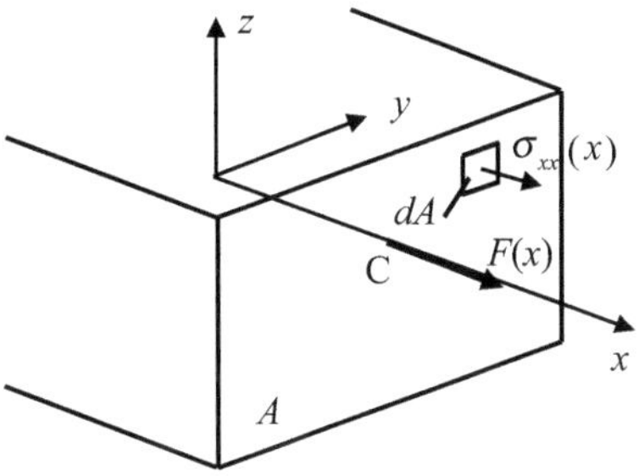

Fig. 4.2 A cross-section of an axially loaded bar

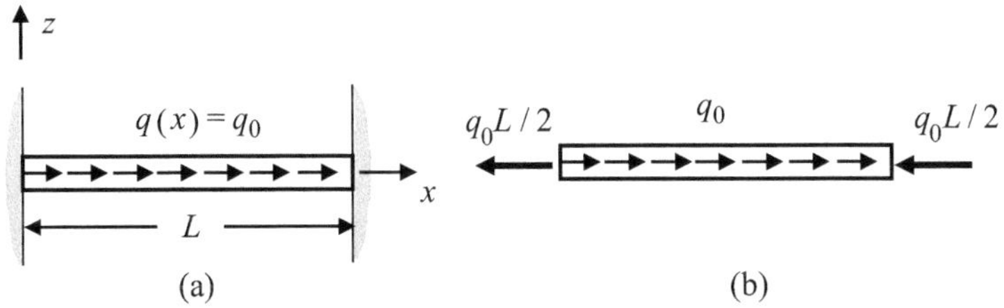

Fig. 4.3 (**a**) A bar fixed at its ends and carrying a uniform distributed force/unit length. (**b**) The reaction forces at the fixed ends

represents the local elongation/unit length at each location in the bar. The stress and strain are related through Young's modulus, E, of the bar as $\sigma_{xx} = E e_{xx}$. For problems where distributed axially loads, $q(x)$, are present, we found previously in Chap. 3 that the equilibrium equation for the internal force F was given by Eq. (3.1), which we repeat here:

$$\frac{dF}{dx} = -q(x) \tag{4.1}$$

Since $F = \sigma_{xx} A = E A e_{xx}$ and $e_{xx} = du/dx$, the force and displacement satisfy:

$$F = EAdu/dx \tag{4.2}$$

and placing Eq. (4.2) into Eq. (4.1), we find

$$EA\frac{d^2u}{dx^2} = -q(x) \tag{4.3}$$

This is the local differential equation of equilibrium for the bar written in terms of the displacements. Unlike Navier's equations, which are the partial differential equilibrium equations of elasticity written in terms of displacements, it is much easier to solve Eq. (4.3) for statically indeterminate problems. Consider, for example, the case shown in Fig 4.3a, where a uniform axial force/unit length $q(x) = q_0$ acts on a bar of length L whose ends are fixed. We can integrate Eq. (4.3) twice to obtain the displacement

$$EAu(x) = -q_0x^2/2 + C_1x + C_2 \tag{4.4}$$

where (C_1, C_2) are constants of integration. Using the conditions that the ends are fixed, namely $u(0) = 0$, $u(L) = 0$, we obtain $C_2 = 0$, $C_1 = q_0L/2$. The internal force in the bar from Eq. (4.2) is

$$F(x) = q_0L/2 - q_0x \tag{4.5}$$

which shows that the reaction forces at the ends are as shown in Fig. 4.3b, a result we could anticipate because of the symmetry of the problem.

4.2.2 A Force-Based Solution

To use the second-generation force-based method on axially loaded bar problems, we need to satisfy conditions analogous to the conditions under which the algebraic equations of linear elasticity were obtained. Recall that (1) the internal stress (and hence the force $F(x)$) is written in terms of discrete force parameters $\{F\}$ and must satisfy homogeneous equilibrium equations, (2) we must express the problem in terms of discrete displacements $\{U\}$ and applied forces, $\{P\}$ acting at discrete locations (called nodes) along the bar, and (3) the bar must have small deformations and behave in a linear elastic manner. To see how to satisfy all these conditions, consider the specific example shown in Fig. 4.4a.

Example 4.1: A Fixed-Fixed Bar—Force-Based Solution
A free body diagram of the bar in Fig. 4.4 (Fig. 4.4b) shows that it is convenient to break it into three elements and define four displacements at the four nodes that define the ends of those elements (the choice of three elements was made only to simplify the problem by making the elements match the loading). The internal force in the middle element will not be the solution of a homogeneous equilibrium equation, where $q(x) = 0$ in Eq. (4.1),

(continued)

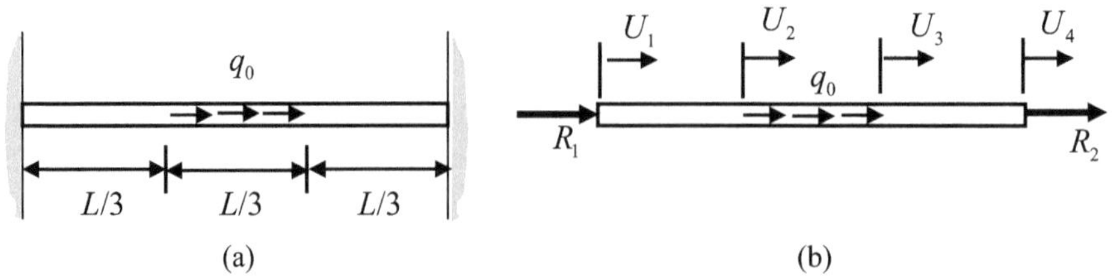

Fig. 4.4 (**a**) A statically indeterminate bar problem. (**b**) A free body diagram of the bar and displacements at four nodes

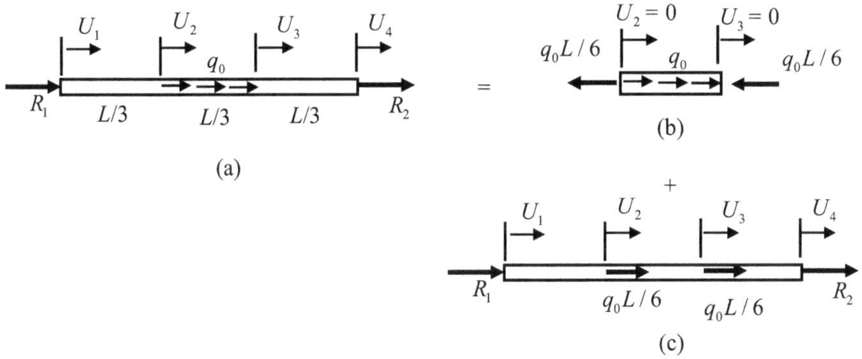

Fig.4.5 (**a**) The forces acting on the bar of Fig. 4.4a. (**b**) A center element whose ends are fixed. (**c**) The bar of Fig. 4.5 where only discrete forces are acting at the nodes where the displacements are defined

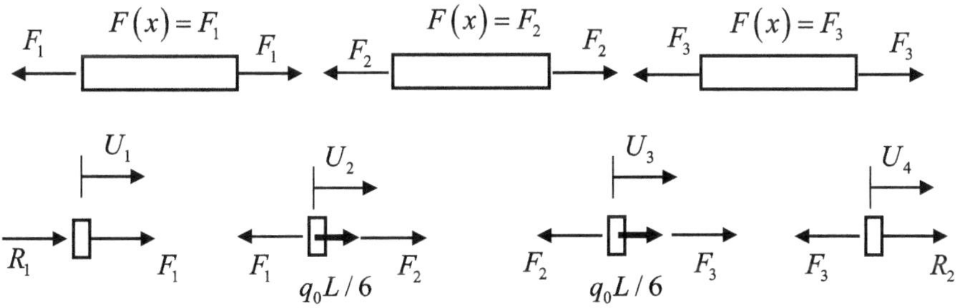

Fig. 4.6 The three elements of the bar in Fig. 4.4a where the forces are obtained from the equilibrium at the four nodes where the displacements are defined

Fig. 4.7 The internal force acting in the bar shown in Fig. 4.4a

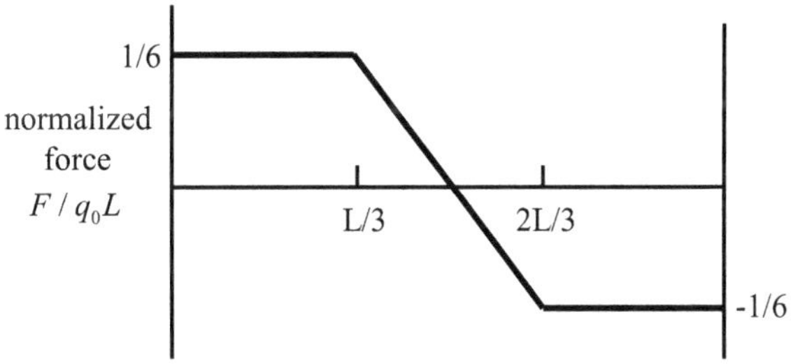

Example 4.1 (continued)
so the conditions under which we can apply the force-based method are not met. However, we can treat the problem shown in Fig. 4.4b as the superposition of the two separate problems seen in Fig. 4.5. The first problem is where

(continued)

Example 4.1 (continued)
we solve the equilibrium of the middle element when its ends are fixed (Fig. 4.5b). We have already found the solution for the end reactions in this problem (see Fig. 4.3b). For distributed forces with different distributions, we can also find the end reactions in the same fashion. If we simply apply the negative of the reactions at nodes 2 and 3, as shown in Fig. 4.5c, then we have a problem where all the conditions for applying the force-based method are satisfied. Solving the problem for Fig. 4.5c and then adding that solution to the solution of Fig. 4.5b give us the solution to our original problem. Since the problem of Fig. 4.5b is solved under fixed-fixed end conditions, the zero displacements at those ends do not affect the displacements for the problem in Fig. 4.5c, even though we are not solving for those displacements.

In the problem shown in Fig. 4.5c, the internal forces are simply constants in each of the elements. This is just like the pin-connected truss problems considered in statics where the forces in all the two-force members are constants and we can solve the problem from equilibrium equations written at all the pins. Here, we can solve for those forces as well as the unknown reactions from the equilibrium equations at the four nodes, as shown in Fig. 4.6. There are four nodes and five unknown forces, so the problem is formally statically indeterminate, but from the symmetry of the problem, we expect $R_1 = R_2 = -q_0L/6$. We will obtain the solution, however, without that symmetry assumption. Here are the equilibrium equations at the four nodes:

$$
\begin{aligned}
{}_{1}\sum F_x &= 0 \quad R_1 + F_1 = 0 \\
{}_{2}\sum F_x &= 0 \quad F_2 - F_1 + q_0L/6 = 0 \\
{}_{3}\sum F_x &= 0 \quad F_3 - F_2 + q_0L/6 = 0 \\
{}_{4}\sum F_x &= 0 \quad R_2 - F_3 = 0
\end{aligned}
\tag{4.6}
$$

and the MATLAB solution:

```
syms R1 R2 F1 F2 F3 q0 L
Eq(1) = R1 + F1 == 0;
Eq(2) = F2 - F1 + q0*L/6 == 0;
Eq(3) = F3 - F2 + q0*L/6 == 0;
Eq(4) = R2 - F3 == 0;
T= [R1 R2 F1 F2 F3];
[E , P] = equationsToMatrix(Eq, T);
flex = [0 0 1 1 1];    % the common L/3EA flexibility is not included
G = diag(flex);
S = null(E).';
```

(continued)

Example 4.1 (continued)

```
Esys = [E; S*G];
Psys = [P; 0];
F = Esys\Psys
F = -(L*q0)/6    % Reaction R1
    -(L*q0)/6    % Reaction R2
     (L*q0)/6    % Internal force F1
      0          % internal force F2
    -(L*q0)/6    % internal force F3
```

In this solution, we divided out the common flexibility *L*/3*EA* of each element. We see the reactions agree with the result deduced from symmetry. The forces in elements 1 and 3 are the exact forces found in the original problem since those forces are constants in those elements. To obtain the exact force in element 2, we need to add in the force from the problem shown in Fig. 4.5b. In terms of the distance x as measured from the left end of the entire bar we, have in element 2

$$F_2(x) = -q_0(x - L/3) + q_0L/6 \tag{4.7}$$

and the distribution of the internal force in the entire bar is shown in Fig. 4.7.

Note that the force-based method as applied here is very similar to a finite element method since we have broken our structure into elements and solved the problem shown in Fig. 4.6. In a finite element solution, there may be thousands or millions of elements and unknowns, so the solution procedure we used for three elements is not practical for such cases and must be changed. We will discuss those changes later in this chapter.

4.2.3 Distributed Forces and Their Discrete Force Equivalents

When the forces are constants in the elements, the displacements will be linear functions of position in the elements. We can write these displacements in terms of the end displacements. For the jth element of length l_j we have

$$u(\xi) = \left(1 - \frac{\xi}{l_j}\right)U_j + \frac{\xi}{l_j}U_{j+1} \tag{4.8}$$

in terms of the distance ξ as measured from the left end of the element. Recall that in Chap. 2 in Eq. (2.10) we expressed the displacement in an element in a matrix-vector form as:

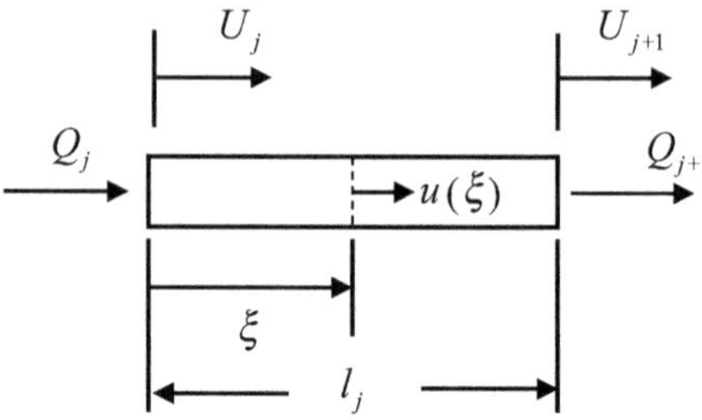

Fig. 4.8 The displacements of an element of the bar. Also shown are the equivalent end forces arising from any distributed forces on the element

$$\{u(\mathbf{x})\} = [H(\mathbf{x})]\{U\} \tag{4.9}$$

and Eq. (4.8) is in this same form where $x = \xi$:

$$u(\xi) = \left[\left(1 - \frac{\xi}{l_j}\right) \quad \frac{\xi}{l_j} \right] \begin{Bmatrix} U_j \\ U_{j+1} \end{Bmatrix} \tag{4.10}$$

and the functions

$$[H(\xi)] = \left[\left(1 - \frac{\xi}{l_j}\right) \quad \frac{\xi}{l_j} \right] \tag{4.11}$$

are called shape functions.

Displacement shape functions give us an alternate way to determine the equivalent forces from distributed loads. In deriving the algebraic equations of linear elasticity, we saw how distributed body forces $\{f\}$ could be replaced by forces acting at discrete nodes through (see Eq. (2.16)):

$$\{Q\} = \int_V [H]^T \{f\} dV \tag{4.12}$$

For our bar problem we can let $\{f\}dV = \{f\}Ad\xi = \{q\}d\xi$ in terms of a distributed force/unit length $\{q\}$ (where $\{q\}$ in the bar problem will be a scalar) and write Eq. (4.12) as

$$\begin{aligned} \{Q\} &= \int_{l_j} [H]^T \{q\} d\xi \\ \rightarrow \begin{Bmatrix} Q_j \\ Q_{j+1} \end{Bmatrix} &= \int_0^{l_j} \begin{bmatrix} 1 - \xi/l_j \\ \xi/l_j \end{bmatrix} q(\xi) d\xi \end{aligned} \tag{4.13}$$

The equivalent end forces acting on an element are shown in Fig. 4.8. For a constant distributed force/unit length $q(\xi) = q_0$ in the element, we can perform the integral in Eq. (4.13) and find the forces $\{Q\}$ as

$$\begin{Bmatrix} Q_j \\ Q_{j+1} \end{Bmatrix} = \begin{Bmatrix} q_0 l_j/2 \\ q_0 l_j/2 \end{Bmatrix} \tag{4.14}$$

and these are precisely the negative of the reactions found by solving for an element with its ends fixed and acted upon by q_0, where in the problem just considered, we had $l_j = L/3$. The end forces $\{Q\}$, therefore, can either be found by solving a fixed-fixed bar problem or by performing the integrations in Eq. (4.13). Those integrations are most easily performed by using MATLAB symbolic integration. The results for different distributions are given in Fig. 4.9. As an example, consider the case (b) in Fig. 4.9 where the distributed force is

$$q(\xi) = \begin{cases} q_0 & 0 < \xi < a \\ 0 & a < x < l \end{cases} \tag{4.15}$$

Letting $\xi = x$ and, $q_0 = q$ the equivalent end forces in MATLAB are found as

```
syms x a q l
Q1 = int((1-x/l)*q, x, [0 a])   % determine equivalent force at left end
                                  and show it
Q1 = -(a*q*(a - 2*l))/(2*l)
Q2 = int((x/l)*q,x, [0 a])      % determine equivalent force at right end
                                  and show it
Q2 = (a^2*q)/(2*l)
```

These results are shown in Fig. 4.9b.

The other cases in Fig. 4.9 can be found in a similar manner. The case (f) in Fig. 4.9 may appear unusual since we are replacing a discrete force in an element by two end forces. This is advantageous to use when, for a desired choice of an element, an applied force does not act at one of the ends. For all the distributed force cases in Fig. 4.9, the negative of the equivalent end forces shown in that figure together with the distributed force produce zero displacements at the ends of the element, although we did not have to enforce those conditions directly. The result for case (f) is best obtained by satisfying those zero displacement end conditions explicitly for a fixed-fixed problem, as done originally for the case shown in Fig. 4.3, since the force P is not a distributed force so one cannot use Eq. (4.13). [Note: There is a way to use Eq. (4.13) if we define what are called generalized functions but that takes us past the technical level usually assumed for a strength of materials course.]

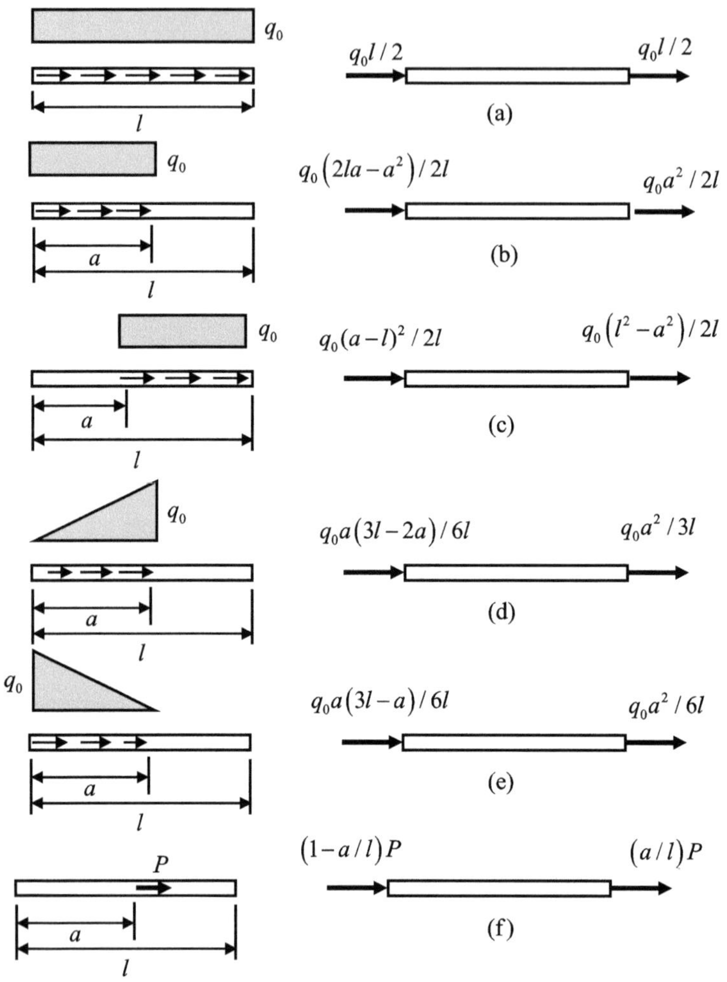

Fig. 4.9 (**a**)–(**e**) Various distributed forces and their equivalent end forces. (**f**) A concentrated applied force and its equivalent end forces

4.2.4 The Force-Based Method for Axial Forces

In this section, we will cover in more detail the general steps that must be followed in solving a statically indeterminate axially loaded bar problem with the force-based method. These steps will be illustrated with the specific example shown in Fig. 4.10a.

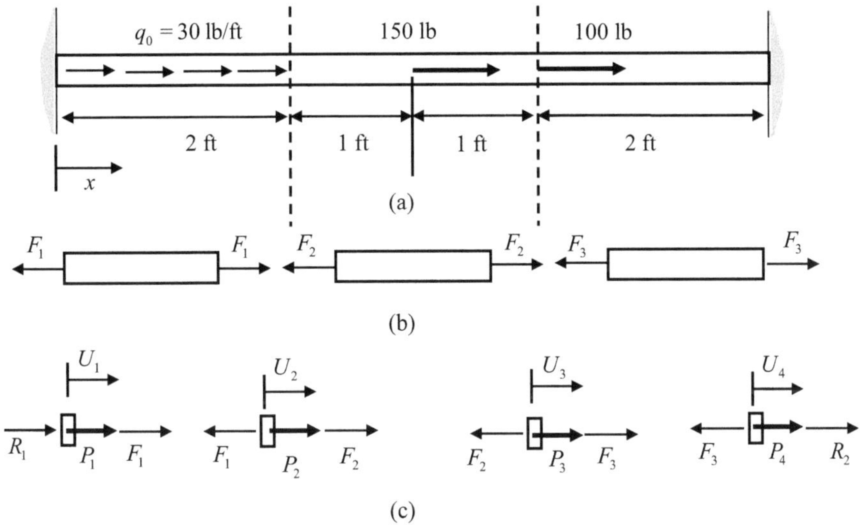

Fig. 4.10 (**a**) A statically indeterminate bar. (**b**) The internal forces in the elements when all the applied forces act at the element ends (nodes). (**c**) Free body diagrams of the forces acting at the nodes

Example 4.2: A Fixed-Fixed Bar—Forces and Displacements

Steps to Determine Forces and Displacements

1. The bar is broken into elements. Suppose, for example, we separate the bar in Fig. 4.10a into three equal length elements. The ends of the elements are called nodes. There are four nodes for the three elements, as shown in Fig. 4.10c.
2. All the applied forces are replaced by equivalent forces acting at the nodes. After that is done, there are no applied forces acting within the elements so that the internal forces in the elements are in equilibrium with themselves, i.e., they satisfy homogeneous equilibrium equations. For axially loaded bars, these internal forces are the constants (F_1, F_2, F_3), as shown in Fig. 4.10b.
3. All the forces acting at the nodes are shown. These include known applied forces at those nodes, if any, and equivalent nodal forces from other forces, such as distributed forces and forces in the element not acting at the nodes. The total forces at the nodes from these known nodal forces or equivalent nodal forces are labeled (P_1, P_2, P_3, P_4) in Fig. 4.10c. There are also unknown reaction forces at the fixed supports. These are the forces (R_1, R_2) in Fig. 4.10c. Finally, there are the internal (F_1, F_2, F_3) forces acting on the nodes from the internal forces. These internal forces are all shown as

(continued)

Example 4.2 (continued)

tensile forces in the elements, so these internal forces will pull on the nodes in the same fashion that tensile truss forces pull on the pins of the truss. The discrete displacements (U_1, U_2, U_3, U_4) at the nodes are also shown in Fig. 4.10c, but we can find the unknown forces directly with our force-based method without knowing those displacements.

4. Equilibrium equations ($\sum F_x = 0$) are written for all the forces at the nodes. Here there are four equilibrium equations and five unknowns:

$$\begin{aligned} {}_1\sum F_x &= 0 \quad F_1 + R_1 + P_1 = 0 \\ {}_2\sum F_x &= 0 \quad F_2 - F_1 + P_2 = 0 \\ {}_3\sum F_x &= 0 \quad F_3 - F_2 + P_3 = 0 \\ {}_4\sum F_x &= 0 \quad R_2 - F_3 + P_4 = 0 \end{aligned} \tag{4.16}$$

which we can write as $[E]\{F\} = \{P\}$ if we let $\{F\} = [F_1 \quad F_2 \quad F_3 \quad R_1 \quad R_2]^T$. In the case shown in Fig. 4.10a, $[E]$ is:

$$[E] = \begin{bmatrix} 1 & 0 & 0 & 1 & 0 \\ -1 & 1 & 0 & 0 & 0 \\ 0 & -1 & 1 & 0 & 0 \\ 0 & 0 & -1 & 0 & 1 \end{bmatrix} \tag{4.17}$$

The column vector $\{P\} = [-P_1 \quad -P_2 \quad -P_3 \quad -P_4]^T$ where the minus signs arise because we have placed the known force on the right side of the equilibrium equations, as we have discussed previously. For the case shown in Fig. 4.10a, we will show shortly that $\{P\} = [-30 \quad -105 \quad -175 \quad 0]^T$.

5. Since the equilibrium matrix is known, we can determine the homogeneous solutions of the equilibrium equations and find the compatibility matrix $[S]$ in the manner described previously. The compatibility matrix satisfies the compatibility equations for the deformations $\{\Delta\}$ given as $[S]\{\Delta\} = \{0\}$ or, in terms of the unknown forces as $[S][G]\{F\} = \{0\}$, where $[G]$ is the flexibility matrix.
6. We need to determine the flexibility matrix $[G]$. It is made up of the flexibilities of the individual elements. For axial force members, the deformation, Δ, and the internal force, F, are related by $\Delta = gF$, where $g = l/AY_m$ is the flexibility, l is the length of the element, A is the cross-sectional area, and Y_m is Young's modulus (Young's modulus is usually written as E, but in that case there is a conflict with using E for the equilibrium matrix).

(continued)

Example 4.2 (continued)

These flexibilities are placed along the diagonal of the flexibility matrix. For non-flexible force such as (R_1, R_2), the flexibilities on the diagonal are set equal to zero. If, for example, in the problem of Fig. 4.10a, the parameters (l, A, Y_m) were the same for each element, the flexibility matrix would be

$$[G] = \frac{l}{AY_m}\begin{bmatrix} 1 & 0 & 0 & 0 & 0 \\ 0 & 1 & 0 & 0 & 0 \\ 0 & 0 & 1 & 0 & 0 \\ 0 & 0 & 0 & 0 & 0 \\ 0 & 0 & 0 & 0 & 0 \end{bmatrix} \tag{4.18}$$

7. We now have all the quantities needed to determine the unknown internal forces and reactions by solving for $\{F\}$ the combination of the equilibrium and compatibility equations:

$$\begin{bmatrix} [E] \\ - \\ [S][G] \end{bmatrix}\{F\} = \begin{Bmatrix} \{P\} \\ - \\ \{0\} \end{Bmatrix} \tag{4.19a}$$

which we have written as

$$[E_{sys}]\{F\} = \{P_{sys}\} \tag{4.19b}$$

This is the solution for the forces. To find the solution for the four nodal displacements shown in Fig. 4.10c, written as the column vector $\{U\} = [U_1 \quad U_2 \quad U_3 \quad U_4]^T$, we must solve the equations (see Eq. (2.34))

$$[E_{sys}]^T\begin{Bmatrix} \{-U\} \\ - \\ \{0\} \end{Bmatrix} = [G]\{F\} \tag{4.20}$$

where in Eq. (4.20) we have indicated explicitly that the negative of the displacements in the x-direction is obtained because vector of known forces $\{P\} = [-P_1 \quad -P_2 \quad -P_3 \quad -P_4]^T$ also contained the negative of the

(continued)

Example 4.2 (continued)
forces in the x-direction. The column vector {0} contains as many zeros as there are compatibility equations. For the problem of Fig. 4.10a, there is only one compatibility equation, so there is only one zero in {0}. Eq. (4.20) only gives the displacements at the nodes. To obtain the displacements for any point in the bar, first note that our solution for $\{F\}$ also contained the reaction forces. If the reactions are known, we can directly determine the internal force $F(x)$ and then integrate that force to obtain $u(x)$. We will show an example of that method shortly.

You should recognize the similarity of this bar solution to the force-based solution obtained for trusses in Chap. 3. There we obtained the equilibrium equations by examining all the forces acting at the pins. Here, after finding equivalent forces acting at the nodes, we applied equilibrium at the nodes. This same procedure can be used for members under torsion and bending so that all the statically indeterminate problems normally treated in a strength of materials course can be solved using extensions of the approach used for trusses in statics.

Consider now in more detail the MATLAB solution of the problem of Fig. 4.10a where each element has a length $l = 2$ ft $= 24$ in, and we will let the cross-sectional area $A = 2$ in^2 and Young's modulus $Y_m = 10 \times 10^6$ lb/in^2. Figure 4.10a is shown again in Fig. 4.11 together with the forces acting at the nodes. Determine the internal force distribution $F(x)$ in the bar and the displacement distribution $u(x)$.

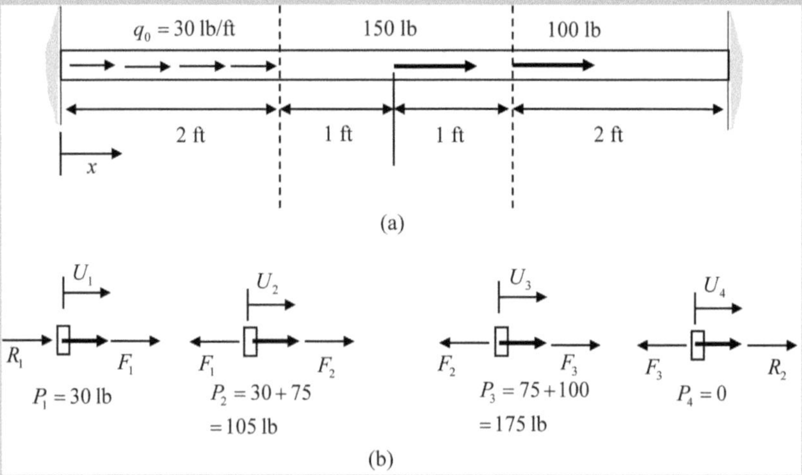

Fig. 4.11 (**a**) The statically indeterminate bar. (**b**) The forces (P_1, P_2, P_3, P_4) which are the sum of the applied and equivalent forces at the nodes

(continued)

Example 4.2 (continued)

First, we need to define the equivalent forces acting at the nodes so that that we can add them to the applied forces already acting at the nodes. For the distributed 30 lb/ft force and the concentrated 150 lb force, we can determine their equivalent nodal forces from Fig. 4.9 and add them to the 100 lb force already acting at node 3 to obtain the total nodal forces shown in Fig. 4.11b. Then the equilibrium equations can be written symbolically in MATLAB:

```
syms F1 F2 F3 R1 R2
T = [ F1 F2 F3 R1 R2];
Eq(1) = F1 + R1 +30 == 0;
Eq(2) = F2 - F1 + 105 == 0;
Eq(3) = F3 -F2 + 175 == 0;
Eq(4) = R2 - F3 == 0;
[E, P] = equationsToMatrix(Eq, T);
```

We show the equilibrium matrix [E]. It is the same as found previously (Eq. (4.17)):

```
E
E =

[ 1, 0, 0, 1, 0]
[-1, 1, 0, 0, 0]
[ 0, -1, 1, 0, 0]
[ 0, 0, -1, 0, 1]
P
```

We also show the force vector {P}. We see it contains the negative of the applied forces acting in the positive x-direction

```
P =

-30
-105
-175
0
```

Now, we find the compatibility matrix [S]

(continued)

Example 4.2 (continued)

```
S = null(E).';
```

Since the compatibility equations are homogeneous equations, we will remove the common l/AY_m flexibility term from the flexibility matrix and write it in normalized form:

```
flex = [1 1 1 0 0];
G = diag(flex);
```

Then, we find the solution for all the unknown forces

```
Esys = [E; S*G];
Psys = [P ; 0];
F = double(Esys\Psys)
F =
  128.3333 % force F1 (lb)
   23.3333 % force F2 (lb)
 -151.6667 % force F3 (lb)
 -158.3333 % reaction force R1 (lb)
 -151.6667 % reaction force R2 (lb)
```

You can verify that the reaction forces are in equilibrium with the applied forces. We can also find the nodal displacements (and a zero). First, we need to place the common l/AY_m flexibility back into the flexibility matrix. The units here must all be consistent.

```
l = 24; A = 2; Ym = 10*10^6; % l in inches, A in in^2, Ym in lb/in^2
G = (l/(A*Ym))*G;
```

Then we can solve for the displacements and convert them to double precision numerical values.

```
Usys = double(Esys.'\G*F) % displacements in inches
Usys =
 1.0e-03 *

     0
 -0.1540
 -0.1820
     0
     0
```

(continued)

Example 4.2 (continued)

These are the negative of the displacements in the x-direction because the force vector $\{P\}$ contained the known forces acting at the nodes in the negative x-direction, as discussed earlier. The displacements (U_1, U_4) are zero as they should be at the fixed supports and there is an additional zero. Thus, the four nodal displacements in the x-direction are:

$$\begin{aligned} U_1 &= 0 \\ U_2 &= 0.154 \times 10^{-3} \text{ in} \\ U_3 &= 0.1820 \times 10^{-3} \text{ in} \\ U_4 &= 0 \end{aligned}$$

4.2.5 Singularity Functions

In the solution for the internal forces (F_1, F_2, F_3) in the problem just considered, only the force F_3 represents the internal force for our original problem since the other two elements also contained other forces. Thus, to obtain the internal force $F(x)$ anywhere in the bar, we need to combine solutions for all the original forces. We did this earlier for the problem of Fig. 4.4a by adding the solutions directly, but this becomes more cumbersome as the number of elements increases. However, we also found the reaction forces as part of the solution, and once those reactions are known all the forces acting on the bar are known so that we can use those forces to write the internal force directly. Since the internal force changes from element to element, normally we must obtain $F(x)$ individually for each element, i.e., in a piecewise fashion for the entire bar. We would like to be able to write the internal force in any element that are in terms of functions valid for the entire bar. We can do this with what are called singularity functions. Consider, for example, a concentrated force P acting at a point $x = a$ anywhere in the bar. The internal force produced by P can be written in piecewise fashion as shown in Fig. 4.12a. Similarly, a constant distributed force that starts at $x = a$ and continues for all $x > a$ can be written in piecewise fashion, as shown in Fig. 4.12b. We could write other similar distributions in the same way. However, if we write these piecewise internal forces expressions as singularity functions (also called Macaulay bracket functions), expressed as $<x - a>^n$, we can add those functions appropriately to obtain the total internal force anywhere in the bar due to the forces and force distributions acting on the bar. These functions can be differentiated and integrated like the ordinary functions $(x - a)^n$ since one can show

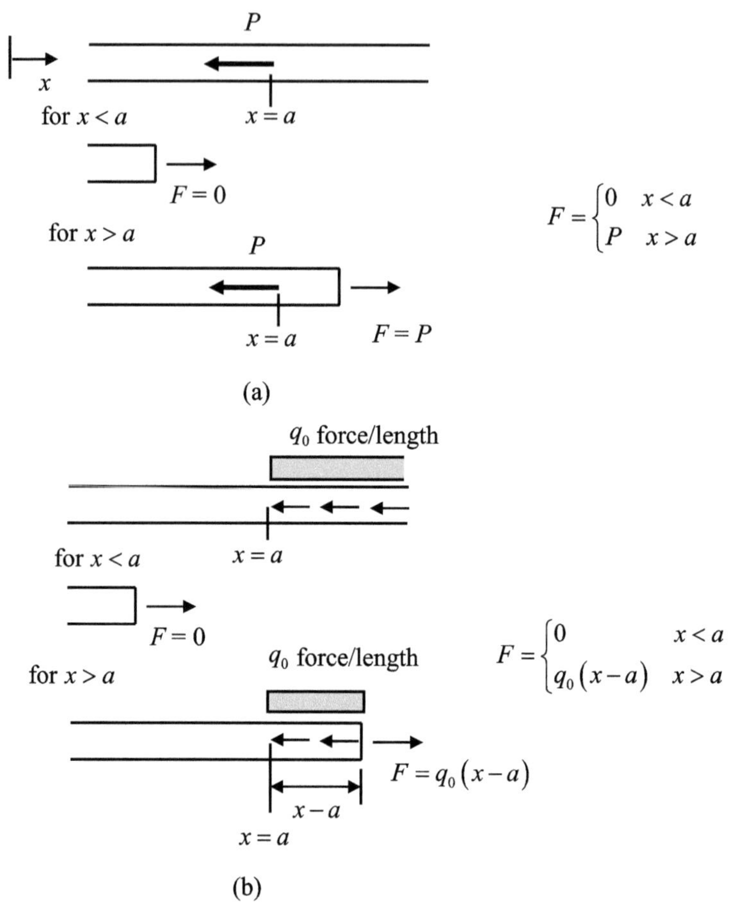

Fig. 4.12 (**a**) A bar where a concentrated force P acts at $x = a$ and the internal force F in the bar. (**b**) A bar where a constant distributed force/unit length begins at $x = a$ and continues indefinitely and the internal force F in the bar

$$\begin{aligned} &\frac{d}{dx}\langle x-a\rangle^{n} = n\langle x-a\rangle^{n-1} \quad n \geq 1 \\ &\int \langle x-a\rangle^{n} dx = \frac{\langle x-a\rangle^{n+1}}{n+1} + C \quad n \geq 0 \end{aligned} \tag{4.21}$$

The internal force expressions in terms of singularity functions for the concentrated force and constant distributed force case are shown in Fig. 4.13 along with the function for a linearly increasing force distribution. A few of these singularity functions have been defined as functions in MATLAB, as shown in Fig. 4.14. The function step_sf models a discontinuous function so if the jump in the function occurs at the first point in the bar, the function needs to be zero at that first point where $x = a$ and then jump to one at the next point to be able to properly plot that

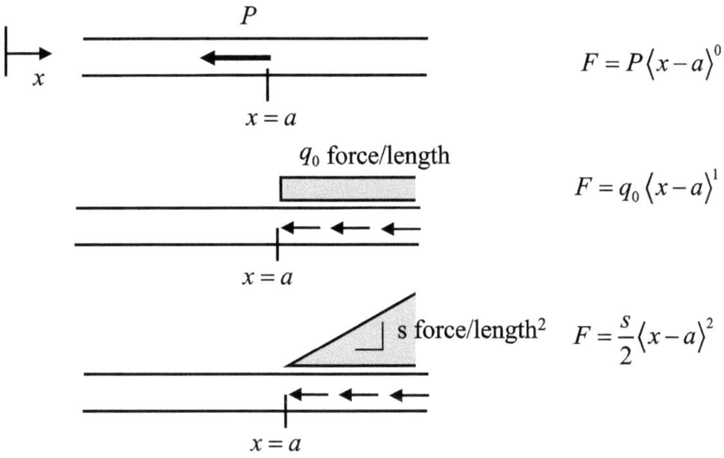

Fig. 4.13 Singularity functions for a few force distributions

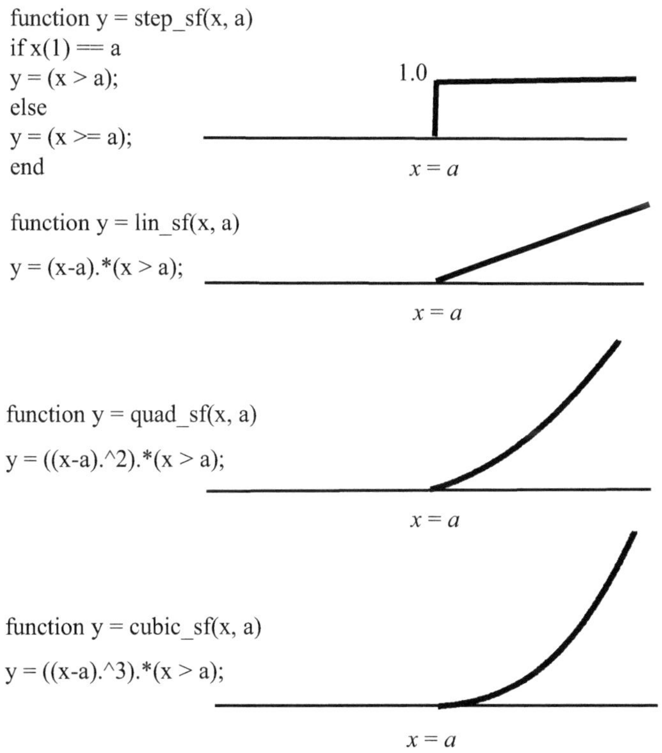

Fig. 4.14 The MATLAB function definitions for some singularity functions and plots of their behavior

jump. At all other points we can take the function to be one at $x >= a$ and the jump will be properly plotted, including at the end point of the bar. The other singularity functions are continuous and equal to zero at $x = a$ so their definitions are simpler.

Example 4.3: A Fixed-Fixed Bar—Use of Singularity Functions
Now, consider the solution of the problem of Fig. 4.10a where the reaction forces were obtained. We will write the solution for the internal force anywhere in the bar and evaluate the internal force at several locations to show the agreement with our force-based solution. We will define an anonymous function in MATLAB which uses all the applied forces and force distributions as well as the reactions:

```
F = @(x) 158.333*step_sf(x,0) -30*lin_sf(x,0) +30*lin_sf(x,2) ...
    -150*step_sf(x,3) -100*step_sf(x,4)+ 151.667*step_sf(x, 6);
```

The step_sf functions represent the discrete applied forces and end reactions. There are two lin_sf functions present because the distributed force starts at $x = 0$ and stops at $x = 2$ ft, so we must use a second singularity function to eliminate the distributed force for $x > 2$ ft. Note the signs present are consistent with the directions shown for the concentrated force and force distributions in Fig. 4.13, which were all shown as acting in the negative x direction. Having defined the internal force, we can now plot it. The results are shown in Fig. 4.15.

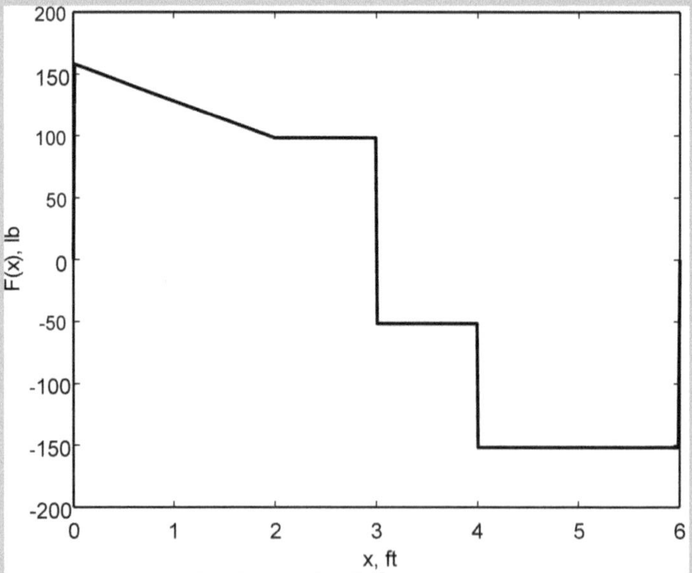

Fig. 4.15 The internal force in the bar of Fig. 4.10a

(continued)

Example 4.3 (continued)

```
x = linspace(0, 6, 1000);
plot(x, F(x))
xlabel(' x, ft')
ylabel('F(x), lb')
```

We see a linearly decreasing internal force in the first element and jumps in the plots at locations where concentrated force act, including the reactions. The force plot starts at zero and ends at zero. If that behavior does not occur, then you know that some forces are wrong in the plot. We can use the anonymous function to evaluate the internal force at any point. For example, in the middle of the third element we find

```
F(5)
ans = -151.6670
```

This is correct because the force in that element is a constant where the force is compressive and equal to the value of the reaction force at $x = 6$ ft. If instead we examine the internal force at $x = 0+$, i.e., just to the right of the fixed support at $x = 0$, we find a tensile internal force whose magnitude is equal to reaction force at that support.

```
F(eps)
ans = 158.3330
```

We can also evaluate the displacements anywhere in the bar since we can use the relationship between the internal force and the displacement (Eq. (4.2)) and integrate that relationship to find

$$u(x) = \frac{1}{Y_m A} \int_0^x F(x)dx \tag{4.22}$$

and there is no constant of integration since the displacement is zero at $x = 0$. We know how to integrate the singularity functions so we obtain the displacement function directly. That function, defined as a MATLAB function, is:

```
A = 2; Ym = 10*10^6;
u = @(x) (12/(Ym*A))*(158.333*lin_sf(x,0) -15*quad_sf(x,0) +
15*quad_sf(x,2) ...
  -150*lin_sf(x,3) -100*lin_sf(x,4) + 151.667*lin_sf(x,6));
```

(continued)

Example 4.3 (continued)

The factor of 12 is present in the function since the distance x is measured in ft, so with that factor, the displacement will be in inches. The plot of the displacement is shown in Fig. 4.16.

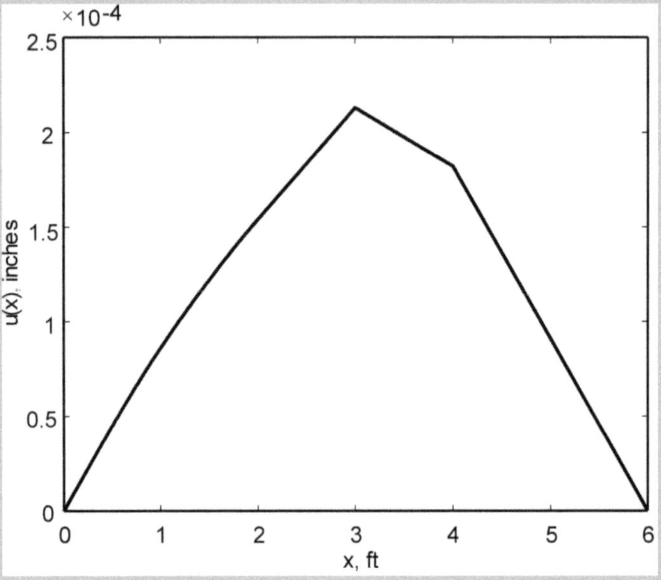

Fig. 4.16 The displacement in the bar of Fig. 4.10a

```
plot(x, u(x))
xlabel(' x, ft')
ylabel('u(x), inches')
axis( [0 6 0 2.5*10^-4])
```

In our solution, we obtained the nodal displacements at $x = 2$ ft and $x = 4$ ft. If we evaluate our anonymous function at these points, we do obtain those nodal displacements:

```
u(2)
ans = 1.5400e-04
u(4)
ans = 1.8200e-04
```

Using singularity functions and MATLAB allows the evaluation of the internal force and displacement anywhere in the bar in a very effective manner.

(continued)

Example 4.3 (continued)
It does require that we define the singularity functions in MATLAB and learn how to construct anonymous functions. There are other less convenient ways to obtain the same results at the level of a strength of materials course without using MATLAB.

4.2.6 Deformation-Displacement Relations

In the solution of the two problems of Figs. 4.4a and 4.10a, the equilibrium equations and the deformation-force relations were used, but the deformation-displacement relations were not described explicitly because they were not needed to obtain the compatibility matrix. However, finding the compatibility matrix from the homogeneous solutions of the equilibrium equations did rely on the assumption that the deformations and displacements are related through the transpose of the equilibrium matrix, i.e., $\{\Delta\} = [E]^T\{U\}$, so let us examine more closely those deformation-displacement relations for the problems of Fig. 4.4a and Fig. 4.10a to guarantee that assumption was valid. Consider first the problem of Fig. 4.4a. The equilibrium equations were given in Eq. (4.6), which in the matrix-vector form $[E]\{F\} = \{P\}$ are

$$\begin{bmatrix} 1 & 0 & 1 & 0 & 0 \\ 0 & 0 & -1 & 1 & 0 \\ 0 & 0 & 0 & -1 & 1 \\ 0 & 1 & 0 & 0 & -1 \end{bmatrix} \begin{Bmatrix} R_1 \\ R_2 \\ F_1 \\ F_2 \\ F_3 \end{Bmatrix} = \begin{Bmatrix} 0 \\ -q_0L/6 \\ -q_0L/6 \\ 0 \end{Bmatrix} \tag{4.23}$$

In this case, the displacements appearing in the deformation-displacement relations are $\{U\} = [-U_1 \quad -U_2 \quad -U_3 \quad -U_4]^T$, where the minus signs are present because the forces in $\{P\}$, as seen in Eq. (4.23), are the negative of the applied forces in the x-direction. Associated with the five forces are five deformations $(\Delta_{R_1}, \Delta_{R_2}, \Delta_1, \Delta_2, \Delta_3)$. If the deformation-displacement relations are in the form $\{\Delta\} = [E]^T\{U\}$, then from Eq. (4.23), we should have

$$\begin{Bmatrix} \Delta_{R_1} \\ \Delta_{R_2} \\ \Delta_1 \\ \Delta_2 \\ \Delta_3 \end{Bmatrix} = \begin{bmatrix} 1 & 0 & 0 & 0 \\ 0 & 0 & 0 & 1 \\ 1 & -1 & 0 & 0 \\ 0 & 1 & -1 & 0 \\ 0 & 0 & 1 & -1 \end{bmatrix} \begin{Bmatrix} -U_1 \\ -U_2 \\ -U_3 \\ -U_4 \end{Bmatrix} \tag{4.24}$$

But we have set the deformations associated with the reaction force to be zero, so if we write out the equations of Eq. (4.24) explicitly under those conditions, we find

$$\begin{aligned} 0 &= -U_1 \\ 0 &= -U_4 \\ \Delta_1 &= U_2 - U_1 \\ \Delta_2 &= U_3 - U_2 \\ \Delta_3 &= U_4 - U_3 \end{aligned} \tag{4.25}$$

By setting the deformations at the end reaction forces to be zero, we have also ensured the displacements at the ends are zero. The other three equations say that the deformations of the three elements are equal to the difference of the end displacements of each element, i.e., the deformations are just the elongations of the individual elements. Thus, identifying the deformations with the elongations, we do have the deformation-displacement relation in the form $\{\Delta\} = [E]^T\{U\}$ satisfied. Now consider the problem of Fig. 4.10a. The equilibrium matrix was given for this problem in Eq. (4.17), and the equilibrium equations were

$$\begin{bmatrix} 1 & 0 & 0 & 1 & 0 \\ -1 & 1 & 0 & 0 & 0 \\ 0 & -1 & 1 & 0 & 0 \\ 0 & 0 & -1 & 0 & 1 \end{bmatrix} \begin{Bmatrix} F_1 \\ F_2 \\ F_3 \\ R_1 \\ R_2 \end{Bmatrix} = \begin{Bmatrix} -30 \\ -105 \\ -175 \\ 0 \end{Bmatrix} \tag{4.26}$$

so if the deformation-displacement equations are in the form $\{\Delta\} = [E]^T\{U\}$,we should have

$$\begin{Bmatrix} \Delta_1 \\ \Delta_2 \\ \Delta_3 \\ \Delta_{R_1} \\ \Delta_{R_2} \end{Bmatrix} = \begin{bmatrix} 1 & -1 & 0 & 0 \\ 0 & 1 & -1 & 0 \\ 0 & 0 & 1 & -1 \\ 1 & 0 & 0 & 0 \\ 0 & 0 & 0 & 1 \end{bmatrix} \begin{Bmatrix} -U_1 \\ -U_2 \\ -U_3 \\ -U_4 \end{Bmatrix} \tag{4.27}$$

which gives explicitly, with the deformations associated with the reaction forces set to zero

$$\begin{aligned} \Delta_1 &= U_2 - U_1 \\ \Delta_2 &= U_3 - U_2 \\ \Delta_3 &= U_4 - U_3 \\ 0 &= -U_1 \\ 0 &= -U_4 \end{aligned} \tag{4.28}$$

Equation (4.28) is again consistent with writing the deformation-displacement relations in the form $\{\Delta\} = [E]^T\{U\}$. Obtaining the compatibility matrix in both problems from homogeneous solutions of the equilibrium equations was, therefore, justified. This form of the deformation-displacement relationship comes from the algebraic equations of elasticity, which is the foundation of all problems solved with second-generation force-based methods, so it is true in general, provided the other assumptions made in deriving the algebraic equations of linear elasticity are satisfied.

4.3 Torsion

4.3.1 Governing Differential Equations

Strength of materials courses also examines the torsion (twisting) of bars having a circular cross-section. We will briefly summarize the governing equations for torsion. Figure 4.1b showed an example of a statically indeterminate torsion problem. The restriction to circular cross-sections is necessary because the deformation assumptions that are made in obtaining the stresses and strains are only valid for circular bars. A bar is under torsion when there are internal twisting moments, $T(x)$, acting in the bar along the axis of the bar, as shown in Fig. 4.17a. These produce a shear stress τ in the bar that acts perpendicular to a line drawn from the center of the

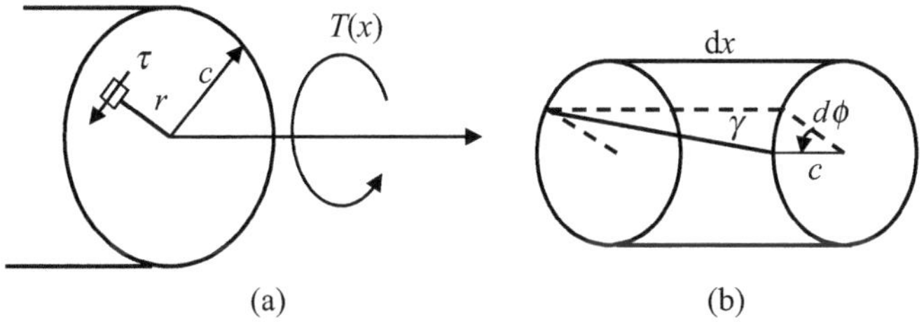

Fig. 4.17 (**a**) The shear stress τ and the internal torque $T(x)$ in a circular bar of radius c. (**b**) The twisting deformation in a small section of a bar, where $d\phi$ is the twist over a length dx and the angle γ is the engineering shear strain in the bar at the surface

cross-section, whose radius is c, as shown. The internal toque causes the bar to twist, producing a deformation shown in Fig. 14b, where a small engineering shear strain on the surface of the bar is the change of angle, γ, seen in that figure and where the twist of the bar over a small length dx is the angle $d\phi$ shown. Thus, the shear strain on the surface is given as $\gamma = cd\phi/dx$. The same deformation is assumed to occur at any radius r so that in general the shear strain is $\gamma = rd\phi/dx$. The shear stress and engineering shear strain for a linearly elastic material are related by $\tau = G\gamma$ where G is the shear modulus [Note: to avoid any confusion with the flexibility matrix, we will call that matrix Gm], so the stress-strain relation is:

$$\tau = G\gamma \tag{4.29}$$

Thus, the shear stress in terms of the twist is given by

$$\tau = G\frac{d\phi}{dx}r \tag{4.30}$$

The shear stress and the internal torque are related by

$$T = \int_A r\tau dA = G\frac{d\phi}{dx}\int_A r^2 dA = GJ\frac{d\phi}{dx} \tag{4.31}$$

where

$$J = \int_A r^2 dA \tag{4.32}$$

is called the polar area moment of the cross-section. For a circular section of radius c, $J = \pi c^4/2$. If one uses Eq. (4.31) in Eq. (4.30), we can write the shear stress in terms of the internal torque as

$$\tau = Tr/J \tag{4.33}$$

which is an expression usually derived in a strength of materials course. If we look at the equilibrium of a small element of the bar (Fig. 4.18), we find

$$\begin{aligned} &\sum M_x = 0 \quad \left(T(x) + \frac{dT}{dx}dx\right) - T(x) + t(x)dx = 0 \\ &\rightarrow \quad \frac{dT}{dx} = -t(x) \end{aligned} \tag{4.34}$$

where $t(x)$ is a distributed applied torque/unit length acting on the bar. Placing Eq. (4.34) into Eq. (4.31) gives

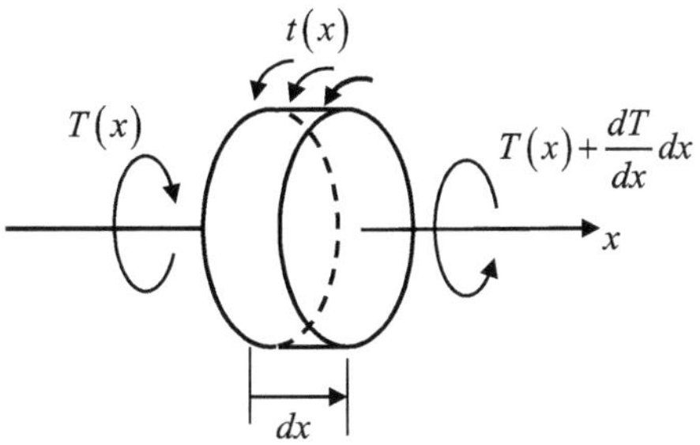

Fig. 4.18 A small element of a circular bar under torsion

$$GJ\frac{d^2\phi}{dx^2} = -t(x) \tag{4.35}$$

which is a differential equation of equilibrium for the bar, written in terms of the angular twist of the bar.

If we examine the governing equations for the twisting of a bar, it follows that uniaxial force problems and torsion problems satisfy very similar equations:

$$\begin{aligned} &\frac{dT}{dx} = -t(x) && \frac{dF}{dx} = -q(x) \\ &T = GJ\frac{d\phi}{dx} && F = EA\frac{du}{dx} \\ &GJ\frac{d^2\phi}{dx^2} = -t(x) && EA\frac{d^2u}{dx^2} = -q(x) \end{aligned} \tag{4.36}$$

so that solutions of either type of problem can be made identical by simply making the substitutions

$$\begin{aligned} &T \leftrightarrow F \\ &\phi \leftrightarrow u \\ &GJ \leftrightarrow EA \\ &t \leftrightarrow q \end{aligned} \tag{4.37}$$

Thus, all the details of solving problems for bars under axial forces can be used for corresponding torsion problems. For example, we can describe distributed and concentrated torques in terms of equivalent end torques, as shown for several cases in Fig. 4.19. The other axial cases seen in Fig. 4.9 also have their torsion counterparts. We can also express concentrated and distributed torques in terms of singularity functions, as seen for several cases in Fig. 4.20.

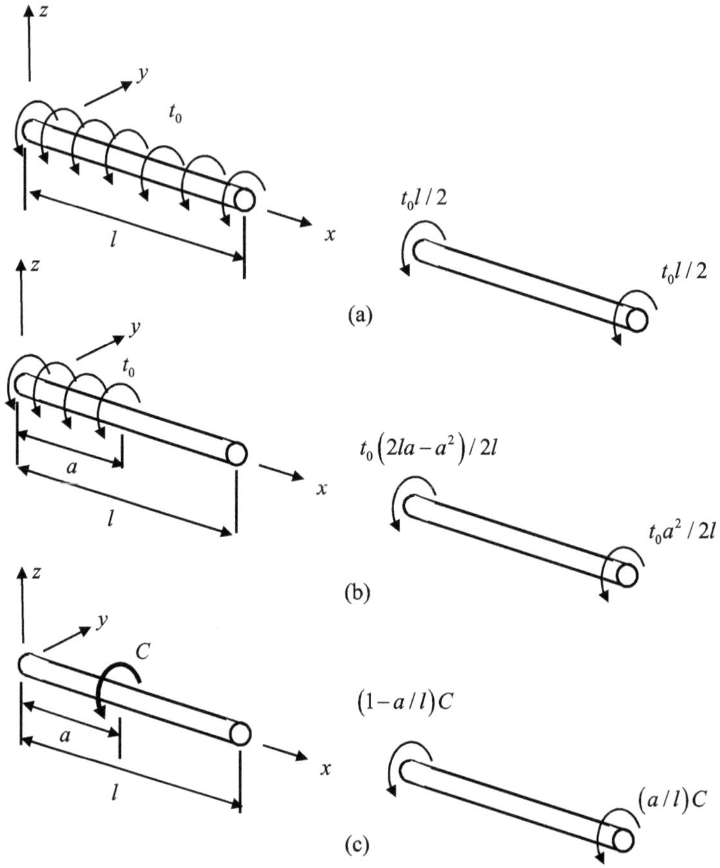

Fig. 4.19 Equivalent end forces for (**a**) a constant distributed torque/unit length, (**b**) a constant torque/unit length only over $0 < x < a$, (**c**) a concentrated torque acting at $x = a$

Fig. 4.20 Singularity functions for (**a**) a concentrated torque at $x = a$, (**b**) a uniform torque/unit length starting at $x = a$

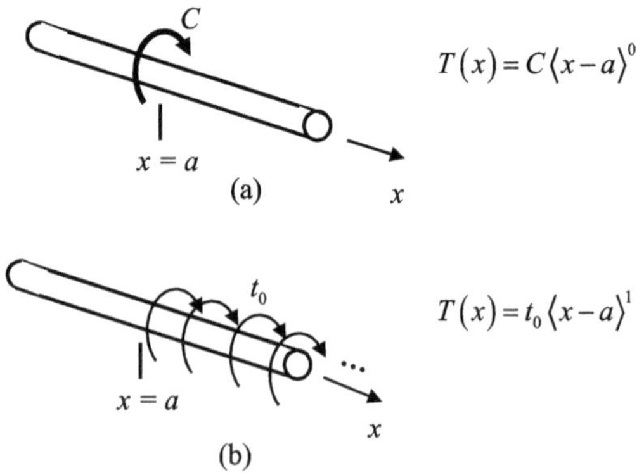

4.3.2 A Force-Based Solution

Example 4.4: A Fixed-Fixed Bar Under Torsion
Now, let's solve the statically indeterminate torsion problem shown in Fig. 4.21a where a circular bar is fixed at both ends and subjected to a constant distributed torque of 400 ft-lb/ft over the first two feet of the bar. To solve this problem, we will use the two elements shown in Fig. 4.21b. The distributed torque has been replaced by the generalized nodal forces (torques) (P_1, P_2). The end reaction torques are (T_{R1}, T_{R2}). Since all the applied torques are at the nodes, the internal torques in the two elements are the constant values (T_1, T_2), respectively. The internal torque $T(x)$ is a constant in each element, so the internal angular displacement (rotation) of the cross section, $\phi(x)$, will be a linear function of x in each element. The rotations at the nodes are the (ϕ_1, ϕ_2, ϕ_3) shown in Fig. 4.21b. In the two elements, we have $T_1 = GJ(\phi_2 - \phi_1)/l = (GJ/l)\Delta_1$ and $T_2 = GJ(\phi_3 - \phi_2)/l = (GJ/l)\Delta_2$, where l is the length of the element. Thus, the deformations (Δ_1, Δ_2) are the differences of rotations at the ends of the elements, i.e., the twists of the elements.

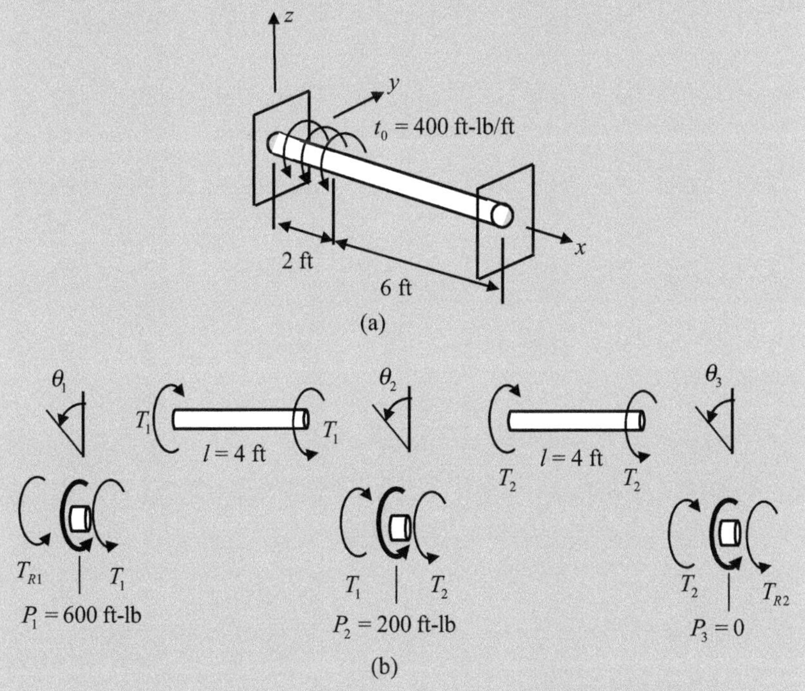

Fig. 4.21 (**a**) A statically indeterminate torsion problem. (**b**) Free body diagrams of the three nodes and the two elements where the distributed torque has been replaced by torques at the nodes

(continued)

Example 4.4 (continued)
Note that for torsion problems, *GJ* plays the same role as *EA* for axial force problems. The torsional flexibility of an element is, therefore, $g = l/GJ$. For this problem, we will assume *GJ* is a constant for the bar. The equilibrium equations at the three nodes are

$$\begin{aligned} {}_1\sum M_x &= 0 \quad T_{R1} + T_1 + 600 = 0 \\ {}_2\sum M_x &= 0 \quad T_2 - T_1 + 200 = 0 \\ {}_3\sum M_x &= 0 \quad T_{R2} - T_2 = 0 \end{aligned} \tag{4.38}$$

Solving this problem for the internal torques with MATLAB gives:

```
syms T1 T2 TR1 TR2 G J l   % G is shear modulus
T = [ T1 T2 TR1 TR2];
Eq(1) = TR1 + T1 + 600 == 0;
Eq(2) = T2 -T1 + 200 == 0;
Eq(3) = TR2 -T2 == 0;
[E, P] = equationsToMatrix(Eq, T);
S = null(E).';
flex = (l/(G*J))*[ 1 1 0 0];    % set flexibilities at supports = 0
Gm = diag(flex);                % Gm is flexibility matrix
Esys = [ E; S*Gm];
M = double(Esys\Psys) % internal torques T1, T2, and reactions TR1,
TR2
M =
  100
 -100
 -700
 -100

x = linspace(0, 8, 500);
T = @(x) 700*step_sf(x,0) -400*lin_sf(x, 0) +400*lin_sf(x,2)
    + 100*step_sf(x,8);
plot(x, T(x))
xlabel('x, ft')
ylabel(' T, ft-lb')
```

(continued)

Example 4.4 (continued)

The solution here was independent of the flexibility l/GJ. We used the reaction torques and singularity functions to plot the internal torque throughout the bar. That plot is shown in Fig. 4.22. The torque varies linearly in the section where the distributed torque is present and is constant in the remainder of the bar. Since $d\phi/dx = T/GJ$, we can determine the angular rotation as a function of x by integrating the internal torque expression:

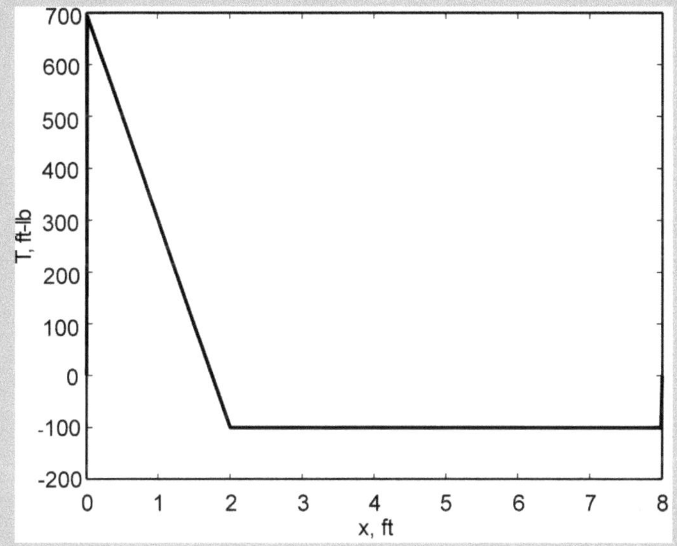

Fig. 4.22 The internal torque for the problem of Fig. 4.21a

```
ur = @(x) 700*lin_sf(x,0) -200*quad_sf(x,0) + 200*quad_sf(x,2)
    + 100*lin_sf(x,8);
    plot(x, ur(x))
```

The angular displacement ur $=GJ\phi$ is shown in Fig. 4.23, where the labels have been added separately. The rotation varies quadratically in the section where the distributed torque is present and varies linearly in the remainder of the bar. The conditions of zero rotation at the ends are clearly satisfied.

(continued)

Example 4.4 (continued)

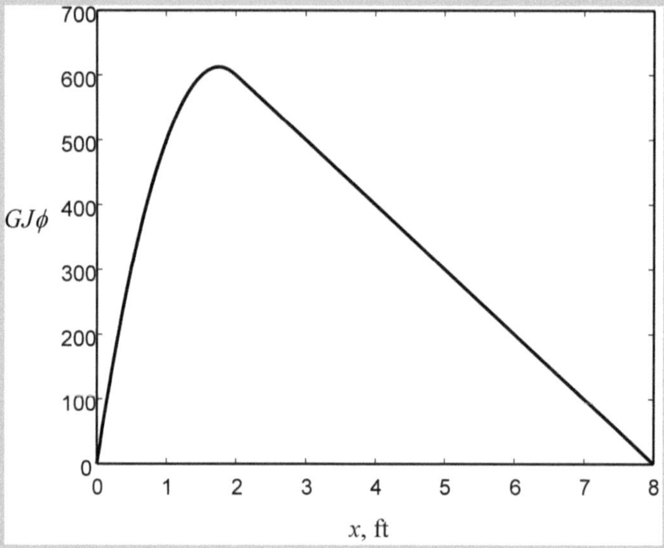

Fig. 4.23 The angular displacement $GJ\phi$ for the problem of Fig. 4.21a

4.4 The Pseudo Stiffness Matrix and Solving for Displacements

Example 4.5: A Fixed-Fixed Bar—Pseudo Stiffness Matrix
As seen previously, the algebraic equations of linear elasticity also give us a new way to introduce the concept of a stiffness matrix and solve problems directly for the displacements. We will use the problem of Fig. 4.10 as an explicit example of this approach. For convenience, Fig. 4.11 is given again as Fig. 4.24. The first steps in obtaining the displacements are the same as in the force-based method, namely we need to replace the distributed forces by equivalent nodal forces and examine the forces and displacements at the nodes, as shown in Fig. 4.24b. From this point, there are two ways to proceed. In the first method, we only look at equilibrium equations at nodes where there are no unknown reaction forces. In this case, these are nodes 2 and 3 and the equilibrium equations at these nodes are (see Eq. (4.16)):

(continued)

Example 4.5 (continued)

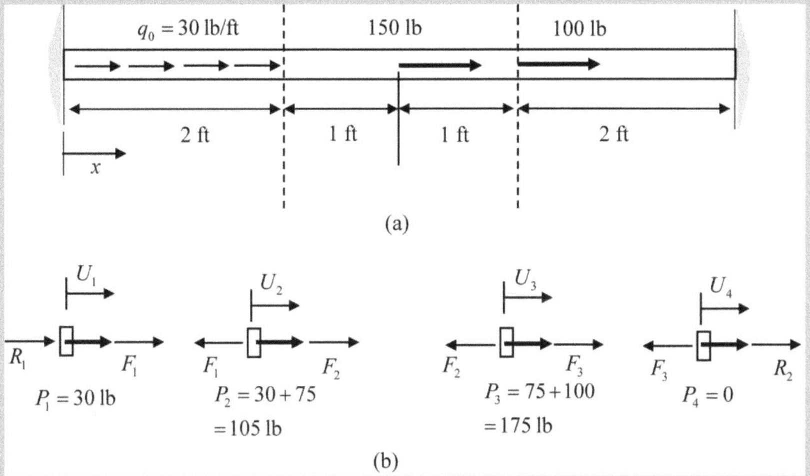

Fig. 4.24 (**a**) The statically indeterminate problem of Fig. 4.10a. (**b**) The displacements and forces at the nodes

$$\begin{aligned} {}_2\sum F_x &= 0 \quad F_2 - F_1 + P_2 = 0 \\ {}_3\sum F_x &= 0 \quad F_3 - F_2 + P_3 = 0 \end{aligned} \tag{4.39}$$

which we will express in matrix-vector form as $[E]\{F\} = \{P\}$. In our present case, we have

$$[E]\{F\} = \{P\} \rightarrow \begin{bmatrix} -1 & 1 & 0 \\ 0 & -1 & 1 \end{bmatrix} \begin{Bmatrix} F_1 \\ F_2 \\ F_3 \end{Bmatrix} = \begin{Bmatrix} -P_2 \\ -P_3 \end{Bmatrix} = \begin{Bmatrix} -105 \\ -175 \end{Bmatrix} \tag{4.40}$$

The deformation-displacement relations are $\{\Delta\} = [E]^T\{U\}$, which in this case gives

$$\{\Delta\} = [E]^T\{U\} \rightarrow \begin{Bmatrix} \Delta_1 \\ \Delta_2 \\ \Delta_3 \end{Bmatrix} = \begin{bmatrix} -1 & 0 \\ 1 & -1 \\ 0 & 1 \end{bmatrix} \begin{Bmatrix} -U_2 \\ -U_3 \end{Bmatrix} \tag{4.41}$$

These are the relationship between the elongations of the elements and the nodal displacements when $U_1 = U_4 = 0$, i.e., we have already satisfied the

(continued)

Example 4.5 (continued)
displacement boundary conditions at the fixed support. The negative signs appear because the vector$\{P\}$ contained the applied forces acting in the negative x-direction. To turn the equilibrium equations into equations for the displacements, we need the deformation-displacement relations expressed in terms of the forces as $\{F\} = [G]^{-1}[E]^T\{U\}$ and then place those relations into the equilibrium equations, which now become:

$$[E][G]^{-1}[E]^T\{U\} = \{P\} \tag{4.42}$$

which can be expressed in terms of a pseudo stiffness matrix, $[K_p]$, as

$$[K_p]\{U\} = \{P\} \tag{4.43}$$

To form up this stiffness matrix, we need the inverse of the flexibility matrix. In the problem of Fig. 4.24a, the length l, cross sectional area A, and Young's Modulus Y_m are taken as the same for all the elements, so the flexibility matrix is

$$[G] = \frac{l}{Y_m A}\begin{bmatrix} 1 & 0 & 0 \\ 0 & 1 & 0 \\ 0 & 0 & 1 \end{bmatrix} \tag{4.44}$$

and its inverse is

$$[G]^{-1} = \frac{Y_m A}{l}\begin{bmatrix} 1 & 0 & 0 \\ 0 & 1 & 0 \\ 0 & 0 & 1 \end{bmatrix} \tag{4.45}$$

Now, let's use the symbolic algebra capabilities of MATLAB to evaluate the stiffness matrix

```
syms Ym A l
E = [-1 1 0; 0 -1 1];
Ginv = (Ym*A/l) * [1 0 0;0 1 0; 0 0 1];
K = E*Ginv*E.'
K =

[(2*A*Ym)/l, -(A*Ym)/l]
[ -(A*Ym)/l, (2*A*Ym)/l]
```

(continued)

Example 4.5 (continued)
which we could write in terms of the elastic constant (stiffness) of a bar $k = Y_m A/l$ as

$$[K_p] = k\begin{bmatrix} 2 & -1 \\ -1 & 2 \end{bmatrix} \tag{4.46}$$

If we solve Eq. (4.43) in MATLAB, letting $l = 24$ inches, $A = 2$ in^2, and Young's modulus $Y_m = 10 \times 10^6$ lb/in^2 we find:

```
l = 24; A = 2; Ym = 10*10^6;
E = [ -1 1 0; 0 -1 1];
P = [-105; -175];
G = (1/(Ym*A))*[1 0 0;0 1 0; 0 0 1];
K = E*inv(G)*E.';
U = K\P % note minus signs
U =
  1.0e-03 *

  -0.1540
  -0.1820
F = inv(G)*E.'*U

F =

 128.3333
 23.3333
 -151.6667
```

The displacements obtained were the negative of the displacements at nodes 2 and 3, as found previously with the force-based method, and the forces in the two members also agree with the previous solution. In this solution, we do not obtain the reaction forces since the equilibrium equations involving those forces were not part of the formulation, but we can easily obtain those reactions from the equilibrium equations at nodes 1 and 4. If we want to include those reactions in the formulation, then the equilibrium equations $[E]\{F\} = \{P\}$ will be based on all four nodes (see Eq. 4.26)):

(continued)

Example 4.5 (continued)

$$\begin{bmatrix} 1 & 0 & 0 & 1 & 0 \\ -1 & 1 & 0 & 0 & 0 \\ 0 & -1 & 1 & 0 & 0 \\ 0 & 0 & -1 & 0 & 1 \end{bmatrix} \begin{Bmatrix} F_1 \\ F_2 \\ F_3 \\ R_1 \\ R_2 \end{Bmatrix} = \begin{Bmatrix} -30 \\ -105 \\ -175 \\ 0 \end{Bmatrix} \tag{4.47}$$

and the corresponding deformation-displacement relations are (see Eq. (4.27)):

$$\begin{Bmatrix} \Delta_1 \\ \Delta_2 \\ \Delta_3 \\ \Delta_{R_1} \\ \Delta_{R_2} \end{Bmatrix} = \begin{bmatrix} 1 & -1 & 0 & 0 \\ 0 & 1 & -1 & 0 \\ 0 & 0 & 1 & -1 \\ 1 & 0 & 0 & 0 \\ 0 & 0 & 0 & 1 \end{bmatrix} \begin{Bmatrix} -U_1 \\ -U_2 \\ -U_3 \\ -U_4 \end{Bmatrix} \tag{4.48}$$

In the force-based method, the flexibilities associated with the reaction forces were set to be zero.

For this problem, therefore, we had (see Eq. (4.18):

$$[G] = \frac{l}{AY_m} \begin{bmatrix} 1 & 0 & 0 & 0 & 0 \\ 0 & 1 & 0 & 0 & 0 \\ 0 & 0 & 1 & 0 & 0 \\ 0 & 0 & 0 & 0 & 0 \\ 0 & 0 & 0 & 0 & 0 \end{bmatrix} \tag{4.49}$$

This means that the deformations associated with those forces will also be zero and, as described previously, in the deformation-displacement relations, Eq. (4.48), this will also mean that the displacements at the fixed supports (U_1 and U_4 in our current problem) are zero, as can be seen from Eq. (4.48). However, in forming up a stiffness matrix, we need to compute the inverse of the flexibility matrix and that inverse does not exist if there are zero flexibilities present. As seen previously, the way to accommodate that problem is to make the flexibilities at the reactions have non-zero values that are much smaller than the other flexibilities so the inverse can be calculated. For example, we could take

(continued)

Example 4.5 (continued)

$$[G]=\frac{l}{AY_m}\begin{bmatrix}1 & 0 & 0 & 0 & 0\\ 0 & 1 & 0 & 0 & 0\\ 0 & 0 & 1 & 0 & 0\\ 0 & 0 & 0 & 10^{-6} & 0\\ 0 & 0 & 0 & 0 & 10^{-6}\end{bmatrix}$$

Thus, the displacements at nodes one and four will not be identically zero, but they will be very small so our solution will be an approximate one, but a very good approximation. Here are the steps in MATLAB:

```
l = 24; A = 2; Ym = 10*10^6;
E = [1 0 0 1 0; -1 1 0 0 0;
0 -1 1 0 0; 0 0 -1 0 1];
P = [ -30; -105; -175 ; 0];
flex = (1/(Ym*A))*ones(1, 5);
G = diag(flex);
G(4,4) = G(4,4)*10^-6; % make flexibilities of reactions very small
G(5,5) = G(5,5)*10^-6;
K = E*inv(G)*E.';
U = K\P % the negative of the displacements again
U =
 1.0e-03 *

 -0.0000
 -0.1540
 -0.1820
 -0.0000

F = inv(G)*E.'*U % the force in the three elements and reactions

F =

 128.3333
 23.3333
 -151.6667
 -158.3333
 -151.6667
```

(continued)

Example 4.5 (continued)

Now we also find the displacements at nodes 1 and 4. These appear to be zero, but if we display these displacements to more decimal places, we see they are very small but non-zero:

```
format long
U
U =
 1.0e-03 *

 -0.000000189999997
 -0.154000187333332
 -0.182000184666668
 -0.000000182000003
```

If we also examine the stiffness matrix to more decimal places, we will find.

```
K =

 1.0e+11 *

 8.3333416666  -0.0000083333             0             0
 -0.0000083333  0.0000166666  -0.0000083333             0
             0 -0.0000083333   0.0000166666 -0.0000083333
             0             0  -0.0000083333  8.3333416666
```

so the equations being solved will be of the form

$$\begin{bmatrix} Nk_{11} & k_{12} & 0 & 0 \\ k_{21} & k_{22} & k_{23} & 0 \\ 0 & k_{32} & k_{33} & k_{34} \\ 0 & 0 & k_{43} & Nk_{44} \end{bmatrix} \begin{Bmatrix} U_1 \\ U_2 \\ U_3 \\ U_4 \end{Bmatrix} = \begin{Bmatrix} -P_1 \\ -P_2 \\ -P_3 \\ -P_4 \end{Bmatrix} \tag{4.50}$$

where N is a large number. Thus, the first and fourth equations, to a good approximation, are $U_1 = 0$ and $U_4 = 0$. Modifying the stiffness matrix in this fashion to satisfy the displacement boundary conditions at fixed supports is commonly used in the displacement-based finite element method. Setting the flexibilities to be zeros in solving for the forces in the force-based approach plays the same role as making the stiffness large but has the advantage over a stiffness approach in that we can set the flexibilities to be identically zero.

4.5 The Connection to Finite Elements

The force-based solutions we have considered in this chapter have much in common with a force-based finite element method. We will consider finite element methods in Chap. 6, but here, we want to briefly discuss the connection between our current approach and finite elements. Let us use the problem of Fig. 4.4a in our discussion. In that problem, we used three elements and after we had replaced the distributed forces with equivalent nodal forces, we examined the forces at all the nodes (Fig. 4.6) which is also shown in Fig. 4.25. We applied the equations of equilibrium at the nodes (see Fig. 4.25b) to form up the equilibrium equations. Thus, this way of treating the problem was on a node-by-node basis. This is a natural extension of the approach used for truss problems in statics where we only needed to consider the nodes (pins) since the element members themselves are inherently in equilibrium. We will continue to use the same approach in the next chapter when we consider beams and frames. If there are relatively few nodes, this is an effective approach where we write each equilibrium equation by hand and then place them symbolically into MATLAB. MATLAB can then be used to put those equilibrium equations into the matrix-vector form $[E]\{F\} = \{P\}$. In this problem, for example, we are led to the equilibrium equations:

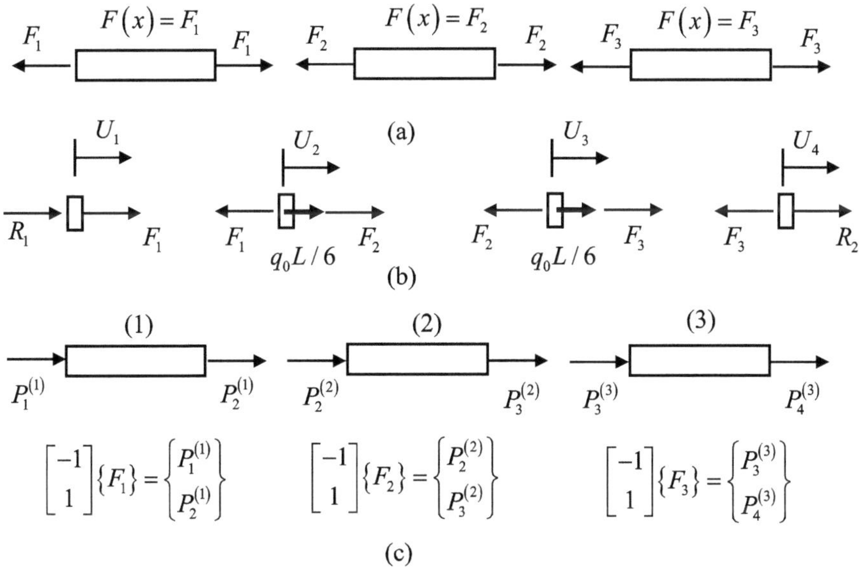

Fig. 4.25 (**a**) The three elements of the bar in Fig. 4.4a. (**b**) Forces and displacements at the four nodes. (**c**) Equilibrium of the three elements, where the internal force in an element is related to the external forces at the ends of the element

$$\begin{bmatrix} 1 & 0 & 1 & 0 & 0 \\ 0 & 0 & -1 & 1 & 0 \\ 0 & 0 & 0 & -1 & 1 \\ 0 & 1 & 0 & 0 & -1 \end{bmatrix} \begin{Bmatrix} R_1 \\ R_2 \\ F_1 \\ F_2 \\ F_3 \end{Bmatrix} = \begin{Bmatrix} 0 \\ -q_0L/6 \\ -q_0L/6 \\ 0 \end{Bmatrix} \tag{4.51}$$

In a finite element approach, we may want to use thousands or even millions of elements so this node-by-node approach is not practical. However, consider forming the equations of equilibrium on an element-by-element basis instead, as shown in Fig. 4.25c. Let the forces acting at the mth node of the (jth) bar element be $P_m^{(j)}$. These end forces are shown in Fig. 4.25c. We want to consider an equilibrium equation for the internal force in each bar element in terms of those nodal forces, so let's begin with an empty 4 × 3 equilibrium matrix

$$\begin{bmatrix} 0 & 0 & 0 \\ 0 & 0 & 0 \\ 0 & 0 & 0 \\ 0 & 0 & 0 \end{bmatrix} \begin{Bmatrix} F_1 \\ F_2 \\ F_3 \end{Bmatrix} = \begin{Bmatrix} 0 \\ 0 \\ 0 \\ 0 \end{Bmatrix} \tag{4.52}$$

For element (1) we have $F_1 = -P_1^{(1)}, F_1 = P_2^{(1)}$ at nodes 1 and 2, respectively, which we could write in matrix-vector form as

$$\begin{bmatrix} -1 \\ 1 \end{bmatrix} \{F_1\} = \begin{Bmatrix} P_1^{(1)} \\ P_2^{(1)} \end{Bmatrix} \tag{4.53}$$

If we place this result into the appropriate places in the matrix of Eq. (4.52), we find

$$\begin{bmatrix} -1 & 0 & 0 \\ 1 & 0 & 0 \\ 0 & 0 & 0 \\ 0 & 0 & 0 \end{bmatrix} \begin{Bmatrix} F_1 \\ F_2 \\ F_3 \end{Bmatrix} = \begin{Bmatrix} P_1^{(1)} \\ P_2^{(1)} \\ 0 \\ 0 \end{Bmatrix} \tag{4.54}$$

Similarly, we have

$$\begin{bmatrix} -1 \\ 1 \end{bmatrix} \{F_2\} = \begin{Bmatrix} P_2^{(2)} \\ P_3^{(2)} \end{Bmatrix}, \quad \begin{bmatrix} -1 \\ 1 \end{bmatrix} \{F_3\} = \begin{Bmatrix} P_3^{(2)} \\ P_4^{(2)} \end{Bmatrix} \tag{4.55}$$

and placing them appropriately into Eq. (4.54), we find for all three elements

$$\begin{bmatrix} -1 & 0 & 0 \\ 1 & -1 & 0 \\ 0 & 1 & -1 \\ 0 & 0 & 1 \end{bmatrix} \begin{Bmatrix} F_1 \\ F_2 \\ F_3 \end{Bmatrix} = \begin{Bmatrix} P_1^{(1)} \\ P_2^{(1)} + P_2^{(2)} \\ P_3^{(2)} + P_3^{(3)} \\ P_4^{(3)} \end{Bmatrix} \tag{4.56}$$

But the forces and sum of forces on the right side of Eq. (4.56) must just be the forces acting at those nodes, so we find

$$\begin{bmatrix} -1 & 0 & 0 \\ 1 & -1 & 0 \\ 0 & 1 & -1 \\ 0 & 0 & 1 \end{bmatrix} \begin{Bmatrix} F_1 \\ F_2 \\ F_3 \end{Bmatrix} = \begin{Bmatrix} R_1 \\ q_0L/6 \\ q_0L/6 \\ R_2 \end{Bmatrix} \tag{4.57}$$

We could remove the first and last equations in Eq. (4.57) to eliminate the unknown reactions and then use the remaining equilibrium equations and a compatibility equation for the internal forces to solve for those forces. But we also could incorporate the reactions into the equilibrium equations by adding two columns to the equilibrium equations:

$$\begin{bmatrix} -1 & 0 & -1 & 0 & 0 \\ 0 & 0 & 1 & -1 & 0 \\ 0 & 0 & 0 & 1 & -1 \\ 0 & -1 & 0 & 0 & 1 \end{bmatrix} \begin{Bmatrix} R_1 \\ R_2 \\ F_1 \\ F_2 \\ F_3 \end{Bmatrix} = \begin{Bmatrix} 0 \\ q_0L/6 \\ q_0L/6 \\ 0 \end{Bmatrix} \tag{4.58}$$

If we compare Eq. (4.58) with Eq. (4.51), we see they are the same equations since both the equilibrium matrix $[E]$ and the vector $\{P\}$ in Eq. (4.57) are the negatives of those appearing in Eq. (4.58). Note that in Eq. (4.58) the vector {P} contains the applied forces acting in the x-direction. Thus, if we write Eq. (4.58) in the form $[E]\{F\} = \{P\}$ and solve these equations, we will obtain the same solution for the forces as found from Eq. (4.51), but when we solve for the displacements, we will find the displacements in the x-direction so we do not have to change their signs.

The procedure we have just illustrated for this problem is the procedure of a force-based finite element solution. A summary of the steps involved is:

Steps in the Force-Based Finite Element Method

1. Obtain the equilibrium matrix for each single element $\left[E_e^{(m)}\right]$. In our case, the element matrices were all the same

(continued)

$$\left[E_e^{(m)}\right] = \begin{bmatrix} -1 \\ 1 \end{bmatrix}$$

2. Assemble those element matrices into a global equilibrium matrix $[E_g]$ for the entire structure.
3. Compute the known nodal forces from any distributed forces present and place all the known forces and the unknown reaction forces acting at the nodes into a global force vector $\{P_g\}$.
4. Either (a) eliminate the equilibrium equations that involve reactions in $\{P_g\}$ and solve for the internal forces from the reduced global equilibrium equations together with the compatibility equations for those reduced equilibrium equations or (b) incorporate the reactions into the global equilibrium equations and solve for internal forces and reactions from that set of equilibrium equations together with the compatibility equations for that global set, setting the flexibilities of the reactions to be zero . In either case, one must compute an appropriate compatibility matrix $[S]$ and have the flexibility matrix $[G^{(m)}]$ for each element assembled into a global flexibility matrix $[G_g]$ for the entire structure so that we can write the compatibility equations in terms of the unknown forces.
5. Solve for the displacements from a knowledge of the forces, as outlined previously.

Computer programs and matrix-vector algebra can implement all these steps, even for very complex problems. Displacement-based finite element methods work in a very similar fashion. Element stiffness matrices are defined and assembled into a global stiffness matrix. The force and displacement boundary conditions are applied to those global equations and then solved for the nodal displacements directly since compatibility equations are not required. Forces are then obtained from the displacements.

Chapter 5
Force-Based Methods in Advanced Strength of Materials

Objectives

- To derive the governing differential equations of beam bending.
- To demonstrate the replacement of distributed forces and couples with equivalent discrete force/moments acting at nodes.
- To apply a second-generation force-based method and displacement-based methods to beam bending problems.
- To consider the extension of force-based methods to frames which can be under combinations of extensional, bending, and torsional deformations.
- To discuss the connection to a force-based finite element method.

5.1 Introduction

An advanced strength of materials course forms a bridge between a strength of materials course and later courses such as elasticity and structural mechanics courses. Typically, it covers more detailed discussions of stresses and strains and presents some elements of the theory of elasticity. Advanced treatments of axial extension, bending, and torsion problems are considered, and work-energy theorems are applied to solving statically indeterminate problems using methods that are like the first-generation force-based method presented in Chap. 1 [1]. In this chapter, we will consider force-based solutions to statically indeterminate beam problems, some examples of which are given in Fig. 5.1. We will also consider frame problems such as shown in Fig. 5.2, which are subjected to combinations of extensional, bending, and torsional deformations. These problems will be solved by breaking a structure into elements and examining the equilibrium of forces and moments at the nodes between the elements, following the same procedures used in previous chapters for truss problems and in the extension and torsion of bars. Equivalent nodal force and

L. W. Schmerr Jr., *Algebraic Equations of Linear Elasticity*,
https://doi.org/10.1007/978-3-031-66174-7_5

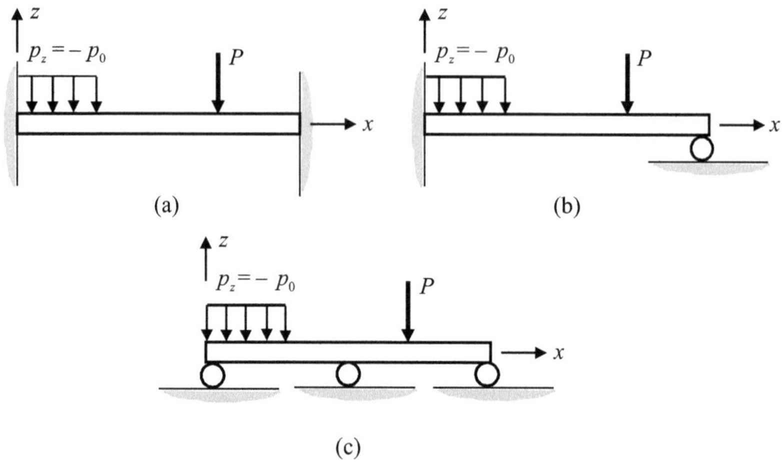

Fig. 5.1 Examples of statically indeterminate beams

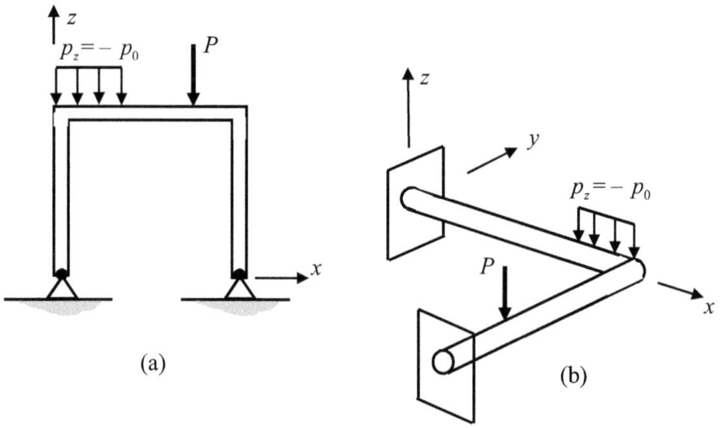

Fig. 5.2 Examples of statically indeterminate frames. (**a**) A pinned frame that is subjected to bending and extension. (**b**) A bent circular bar that is subjected to a combination of bending and torsion

moments will be defined for distributed forces and moments so that all the forces/moments will act on the nodes. There are many aspects of this approach for beams and frames that correspond to obtaining a solution with finite elements so, as done in Chap. 4, we will discuss the connection to a force-based finite element method.

5.2 Beam Bending

5.2.1 Governing Differential Eqs

A beam normally supports vertical loads that may be discrete or distributed, as shown in Fig. 5.1. The internal bending moment in the beam causes fibers along the bottom of the beam such as EA in Fig. 5.3a to be lengthened and fibers along the top of the beam such as DB to be shortened. Thus, there is some fiber in the cross-section that is not stretched or shortened. This is called the neutral axis. It can be shown that the neutral axis is located at the centroid of the cross-section [1]. If we let $w(x)$ be the displacement of the neutral axis in the z-direction then $dw/\mathrm{d}x$ is the slope of that axis, as shown in Fig. 5.3a, which is assumed to be small. If a line that is perpendicular to the neutral axis remains perpendicular to that axis when the beam deforms that line rotates through a small angle ψ which is equal to dw/dx. Thus, at a distance z above the neutral axis the beam will experience a displacement, u_x, in the x-direction (see Fig. 5.3b) given by

$$u_x = -z\frac{dw}{dx} \tag{5.1}$$

which produces a normal strain in the x-direction given by

$$e_{xx} = \frac{\partial u_x}{\partial x} = -z\frac{d^2w}{dx^2} \tag{5.2}$$

This strain produces a normal stress in the x-direction given by

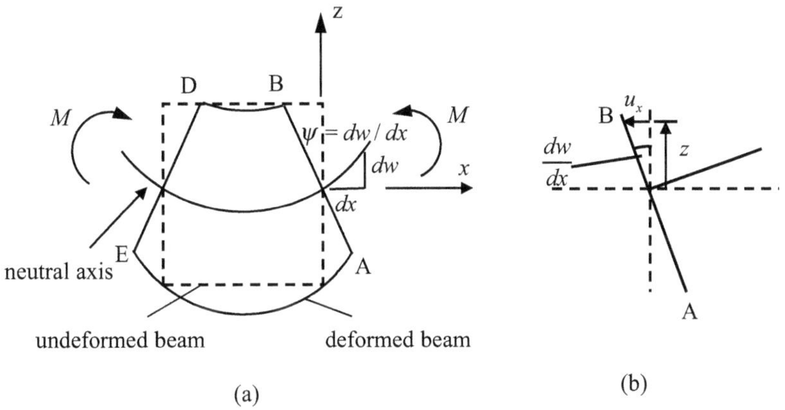

Fig. 5.3 (**a**) The deformation of a beam under bending. (**b**) A diagram of the changes in geometry

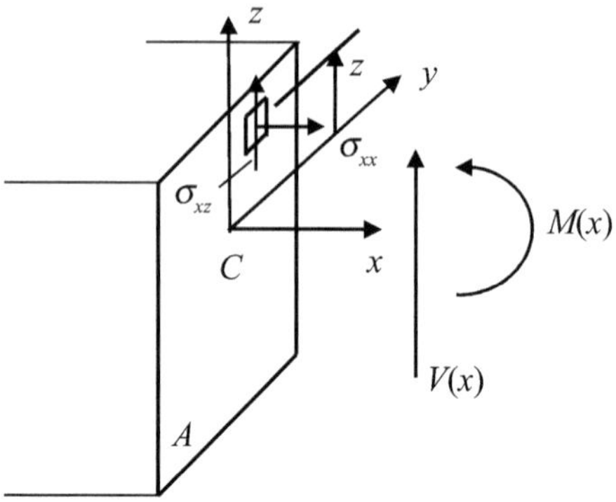

Fig. 5.4 A cross-section of a beam, showing the stresses acting on that cross-section and the internal shear force and bending moment

$$\sigma_{xx} = Ee_{xx} = -zE\frac{d^2w}{dx^2} \tag{5.3}$$

This varying normal stress produces the internal bending moment, $M(x)$ in the cross-section (see Fig. 5.4. The cross-section is shown as rectangular, but the beam can have other shapes. The assumptions made on the deformations are still true for other cross-sections where the z-axis or y-axis is an axis of symmetry [1]). There is also a shear stress σ_{xz} in the cross-section that produces the internal shear force, $V(x)$. These stresses are related to the internal force and moment by integrals:

$$\begin{aligned} M &= -\int_A \sigma_{xx} z dA = E\frac{d^2w}{dx^2}\int_A z^2 dA = EI\frac{d^2w}{dx^2} \\ V &= \int_A \sigma_{xz} dA \end{aligned} \tag{5.4}$$

where

$$I = I_{yy} = \int_A z^2 dA \tag{5.5}$$

is called the second area moment about the y-axis (since the distance z is measured from the y-axis). Thus, using Eq. (5.4), we can also write the normal stress as

$$\sigma_{xx} = -\frac{Mz}{I} \tag{5.6}$$

which is an expression commonly derived in a strength of materials course. If we look at a small section of the beam (Fig. 5.5), the equilibrium of moments about the point O in the negative y-direction gives

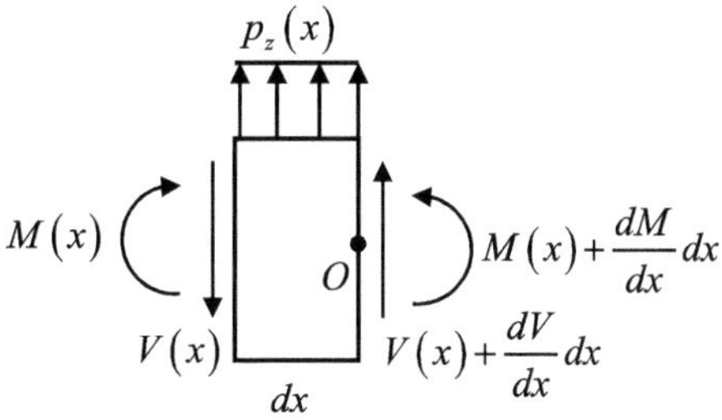

Fig. 5.5 A small element of a beam under bending

$$\left(M(x)+\frac{dM}{dx}dx\right)-M(x)+V(x)dx-p_z(x)dx\left(\frac{dx}{2}\right)=0 \tag{5.7}$$

where $p_z(x)$ is a distributed force/unit length acting on the beam in the z-direction. For a small section dx this distributed force is essentially a constant, as shown in Fig. 5.5. Neglecting the higher order term from p_z in Eq. (5.7), we find

$$\frac{dM}{dx}=-V(x) \tag{5.8}$$

Summing forces in the z-direction, from Fig. 4.21, we also find

$$\begin{aligned}&\left(V(x)+\frac{dV}{dx}dx\right)-V(x)+p_z(x)dx=0\\ \rightarrow\quad &\frac{dV}{dx}=-p_z(x)\end{aligned} \tag{5.9}$$

From Eqs. (5.8) and (5.9), it follows that

$$\frac{d^2M}{dx^2}=p_z(x) \tag{5.10}$$

and from Eqs. (5.10) and (5.4), it follows that

$$EI\frac{d^4w}{dx^4}=p_z(x) \tag{5.11}$$

In summary, for the beam, we have

$$\begin{aligned}
M(x) &= EI\frac{d^2w}{dx^2} \\
\frac{d^2M}{dx^2} &= p_z(x) \\
\frac{dM}{dx} &= -V(x) \\
EI\frac{d^4w}{dx^4} &= p_z(x)
\end{aligned} \tag{5.12}$$

Thus, from force equilibrium in the z-direction and the moment equilibrium about the negative y-axis, we can write the beam equilibrium equations:

$$\begin{aligned}
\frac{d^2M}{dx^2} &= p_z(x) \\
\frac{dV}{dx} &= -p_z(x)
\end{aligned} \tag{5.13}$$

In terms of the vertical displacement of the beam, the corresponding equilibrium is the last equation in Eq. (5.12):

$$EI\frac{d^4w}{dx^4} = p_z(x) \tag{5.14}$$

Solutions of the homogeneous ($p_z = 0$) force and moment equilibrium equations are

$$\begin{aligned}
M(x) &= ax + b \\
V(x) &= -a
\end{aligned} \tag{5.15}$$

where (a, b) are constants. We can choose these constants in a variety of ways. For example, we could let $-a$ be a constant shear force value, V_0, in the beam and the constant b to be the value of the moment at $x = 0$, that is, $b = M(0) \equiv M_1$, in which case we have

$$\begin{aligned}
M(x) &= M_1 - V_0 x \\
V(x) &= V_0
\end{aligned} \tag{5.16}$$

However, because of the important role that beam bending moments play in the deformation of the beam, we will instead choose two discrete moment parameters and set $M_1 = M(0)$, $M_2 = M(l)$, where l is the length of an element. In that case, $a = (M_2 - M_1)/l$, $b = M_1$ and we can write the bending moment and shear force as

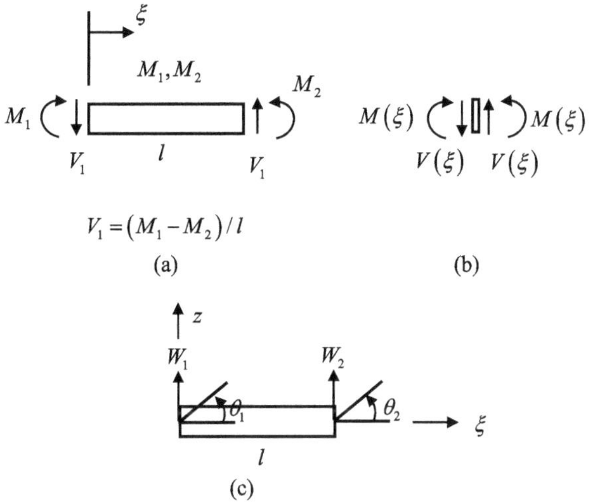

Fig. 5.6 (**a**) The forces and moments on the ends of an element where the internal shear force and bending moment satisfy homogeneous equilibrium equations. (**b**) The internal forces and moments of an infinitesimally small element in the beam. (**c**) The vertical displacements and angular displacements (rotations) at the end nodes

$$\begin{aligned} M(\xi) &= M_1 + (M_2 - M_1)(\xi/l) \\ V(\xi) &= (M_1 - M_2)/l \end{aligned} \tag{5.17}$$

where ξ is the x-distance as measured from the start of an element (see Fig. 5.6a). These expressions make the free body diagrams rather cluttered, so to reduce the complexity, let the constant force in the element be written as V_1, where $V_1 = (M_1 - M_2)/l$. Then the forces/moments at the ends of the element are as shown in Fig. 5.6a. The directions of the forces/moments shown at the ends of the element in Fig. 5.6a are different because they act on different (i.e., back and front) faces. This is also true for the forces and moments on an infinitesimally small section of the beam, as shown in Fig. 5.6b, since from equilibrium of this section, the net force and moment must be zero.

Now consider the displacements for homogeneous solutions of the differential equilibrium equations, Eq. (5.14). The general solution for the z-displacement, $w(\xi)$, of the neutral axis when $p_z = 0$ is

$$w(\xi) = a_1 + a_2\xi + a_3\xi^2 + a_4\xi^3 \tag{5.18}$$

The nodal displacements in the element of the beam shown in Fig. 5.6c are defined by the z-displacements at the nodes (W_1, W_2) and the nodal angular displacements (θ_1, θ_2). Since the slopes of the beam are normally small, they are equal to these angular displacements, that is,

$$\theta_1 = \left.\frac{dw}{d\xi}\right|_{\xi=0}, \quad \theta_2 = \left.\frac{dw}{d\xi}\right|_{\xi=l} \tag{5.19}$$

We can relate the constants in Eq. (5.18) to these nodal displacements by using the conditions

$$\begin{aligned} w(0) &= W_1, \quad w(l) = W_2 \\ \left.\frac{dw}{d\xi}\right|_{\xi=0} &= \theta_1, \quad \left.\frac{dw}{d\xi}\right|_{\xi=l} = \theta_2 \end{aligned} \tag{5.20}$$

These conditions lead to the equations

$$\begin{Bmatrix} W_1 \\ \theta_1 \\ W_2 \\ \theta_2 \end{Bmatrix} = \begin{bmatrix} 1 & 0 & 0 & 0 \\ 1 & 1 & 0 & 0 \\ 1 & l & l^2 & l^3 \\ 1 & 1 & 2l & 3l^2 \end{bmatrix} \begin{Bmatrix} a_1 \\ a_2 \\ a_3 \\ a_4 \end{Bmatrix} \tag{5.21}$$

which can be inverted to give

$$\begin{Bmatrix} a_1 \\ a_2 \\ a_3 \\ a_4 \end{Bmatrix} = \frac{1}{l^3} \begin{bmatrix} l^3 & 0 & 0 & 0 \\ 0 & l^3 & 0 & 0 \\ -3l & -2l^2 & 3l & -l^2 \\ 2 & l & -2 & l \end{bmatrix} \begin{Bmatrix} W_1 \\ \theta_1 \\ W_2 \\ \theta_2 \end{Bmatrix} \tag{5.22}$$

and, when placed into Eq. (5.18), yields

$$\begin{aligned} w(\xi) &= \left(1 - 3\frac{\xi^2}{l^2} + 2\frac{\xi^3}{l^3}\right) W_1 + \left(\xi - 2\frac{\xi^2}{l} + \frac{\xi^3}{l^2}\right)\theta_1 \\ &+ \left(3\frac{\xi^2}{l^2} - 2\frac{\xi^3}{l^3}\right) W_2 + \left(-\frac{\xi^2}{l} + \frac{\xi^3}{l^2}\right)\theta_2 \end{aligned} \tag{5.23}$$

The displacement in the element can be written in matrix vector form as

$$w(\xi) = [H]\{U\}, \quad \{U\} = \begin{Bmatrix} W_1 \\ \theta_1 \\ W_2 \\ \theta_2 \end{Bmatrix} \tag{5.24}$$

where $\{U\}$ is a vector of generalized nodal displacements and $[H]$ is a matrix of shape functions. Although our interest is in solving for the generalized forces such as (M_1, M_2) in the beam we can use these displacement shape functions to define equivalent applied nodal generalized forces, like what was done in axial force and torsional problems. We will discuss this more shortly.

5.2.2 A Force-Based Beam Solution

Example 5.1: A Statically Indeterminate Beam

To see how we can solve beam problems directly for the generalized forces ($M(x)$, $V(x)$) consider the case shown in Fig. 5.7a, where a beam is fixed at its left end and sits on a roller on the right end. The beam carries a uniform distributed force/unit length over its length. This problem can be solved in a manner very similar to what was done in axial force problems. We will consider the original problem as the superposition of the two problems shown in Fig. 5.7b, c. The problem in Fig. 5.7b is a beam of length L which carries the distributed force and is fixed at both its ends (displacements and slopes are both zero at the ends). The reaction forces at ends A and B are the forces/moments (V_A, M_A) and (V_B, M_B). The problem of Fig. 5.7c is the original beam where the negative of the end forces/moments shown in Fig. 5.7b are applied at A and B, as shown, together with the other reaction forces/moments at A and B given by (M_{R1}, V_{R1}) and V_{R2}, respectively. The problem in Fig. 5.7c now is a beam where all the forces/moments act on the nodes at the end of the element (which in this case is the entire beam) and the internal forces/moments in the element will be of the form given in Eq. (5.17) since they satisfy homogeneous equilibrium equations where p_z is zero.

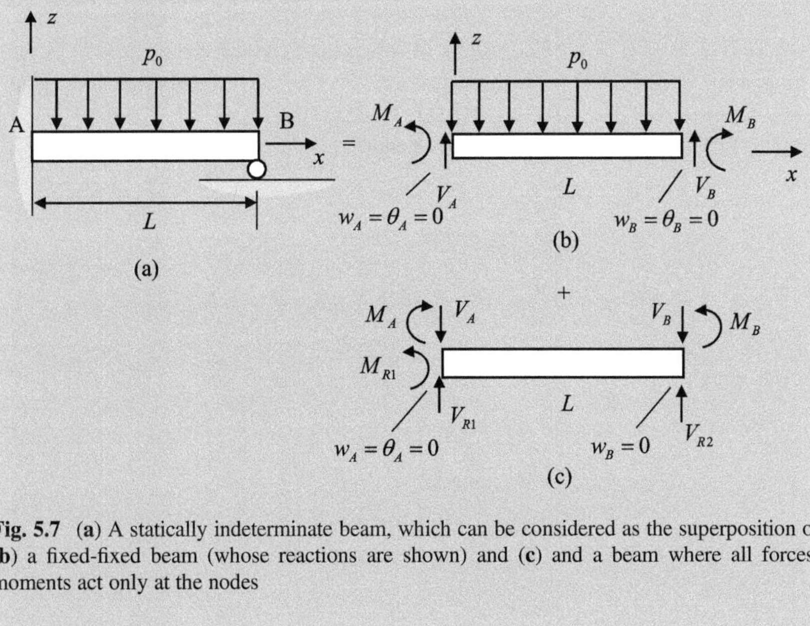

Fig. 5.7 (**a**) A statically indeterminate beam, which can be considered as the superposition of (**b**) a fixed-fixed beam (whose reactions are shown) and (**c**) and a beam where all forces/moments act only at the nodes

(continued)

Example 5.1 (continued)

[Note that the fixed-fixed conditions chosen for the problem of Fig. 5.7b ensure that the generalized displacements (displacements and rotations) at the nodes in the problem of Fig. 5.7c will be unaffected. In a displacement-based approach, we can then solve the problem of Fig. 5.7c for the true nodal generalized displacements. In a force-based method, the replacement of the distributed forces by nodal forces/moments also guarantees that the shear force and bending moments in the elements are solutions of homogeneous equilibrium equations, a key assumption in the force-based approach.]

The first step is to solve the problem shown in Fig. 5.7b. We can integrate Eq. (5.14) four times to find

$$
\begin{aligned}
EI\frac{d^4w(x)}{dx^4} &= -p_0 \\
EI\frac{d^3w(x)}{dx^3} &= -p_0x + C_1 \\
EI\frac{d^2w(x)}{dx^2} &= -p_0\frac{x^2}{2} + C_1x + C_2 \\
EI\frac{dw(x)}{dx} &= -p_0\frac{x^3}{6} + C_1\frac{x^2}{2} + C_2x + C_3 \\
EIw(x) &= -p_0\frac{x^4}{24} + C_1\frac{x^3}{6} + C_2\frac{x^2}{2} + C_3x + C_4
\end{aligned}
\tag{5.25}
$$

where (C_1, C_2, C_3, C_4) are constants of integration. At the fixed end $x = 0$, we have,$w(0) = dw(0)/dx = 0$ which gives $C_3 = C_4 = 0$. At the other fixed end $x = L$, we also have $w(L) = dw(L)/dx = 0$, giving $C_1 = p_0L/2$ and $C_2 = -p_0L^2/12$. From these results, it follows that

$$
\begin{aligned}
V(x) &= -EI\frac{d^3w}{dx^3} = p_0x - p_0L/2 \\
M(x) &= EI\frac{d^2w}{dx^2} = -p_0x^2/2 + p_0Lx/2 - p_0L^2/12
\end{aligned}
\tag{5.26}
$$

from which it follows that the reactions are (see Fig. 5.7b)

$$
\begin{aligned}
V_A &= V(0) = -p_0L/2 \\
M_A &= M(0) = -p_0L^2/12 \\
V_B &= -V(L) = -p_0L/2 \\
M_B &= -M(L) = p_0L^2/12
\end{aligned}
\tag{5.27}
$$

(continued)

Example 5.1 (continued)

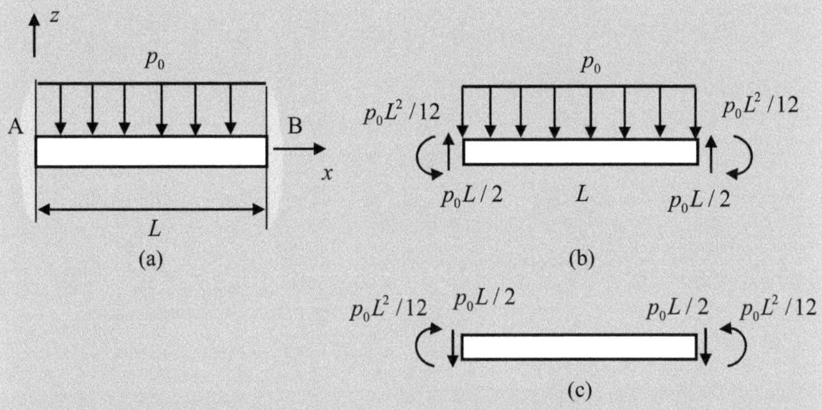

Fig. 5.8 (**a**) A fixed-fixed beam with a uniform distributed force. (**b**) The end reactions. (**c**) The equivalent discrete forces/moments

These reactions for the fixed-fixed problem of Fig. 5.8 (a), therefore, are given explicitly in Fig. 5.8b, and the equivalent discrete reactions from the distributed load are shown in Fig. 5.8c.

The second step is to formulate the equilibrium equations. Figure 5.9a again shows the original problem of Fig. 5.7a we want to solve. To use the force-based method the internal bending moment and shear force must defined in terms of the parameters (M_1, M_2), as shown in Fig. 5.9b. Figure 5.9c shows all the forces/moments acting at the end nodes A and B after the distributed load has been replaced by equivalent forces/moments acting at the end nodes. Those forces/moments include the reactions at the supports and the internal forces and moments acting on the nodes in terms of (M_1, M_2) We can then use Fig. 5.9c to set up the equilibrium equations at those nodes. They are

$$\begin{aligned}
{}_A\sum F_z &= 0 \quad V_{R1} + (M_1 - M_2)/L - p_0L/2 = 0 \\
{}_A\sum M_{-y} &= 0 \quad M_{R1} + M_1 - p_0L^2/12 = 0 \\
{}_B\sum F_z &= 0 \quad V_{R2} - (M_1 - M_2)/L - p_0L/2 = 0 \\
{}_B\sum M_{-y} &= 0 \quad -M_2 + p_0L^2/12 = 0
\end{aligned} \tag{5.28}$$

[Note: Positive moments are taken to be acting about the negative y-axis in the moment equations. Thus, a counterclockwise moment M will be positive when acting in this direction and a counterclockwise beam rotation (slope) θ

(continued)

Example 5.1 (continued)

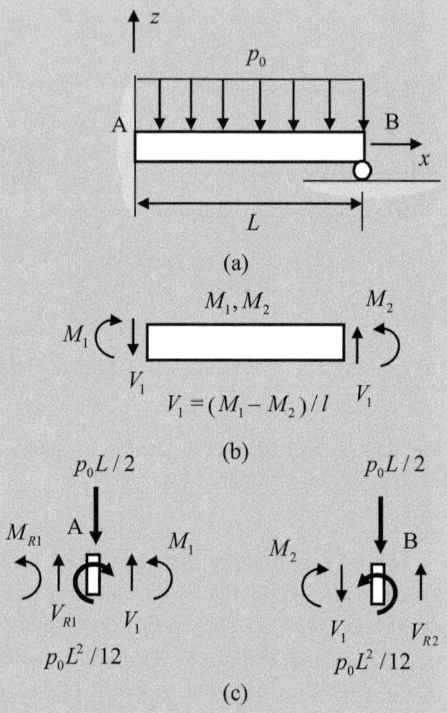

Fig. 5.9 (**a**) The statically indeterminate problem of Fig. 5.7a. (**b**) Bending moment and shear force in the beam defined in terms of the parameters (M_1, M_2). (**c**) All the forces/moments acting at the end nodes when the distributed load has been replaced by equivalent end forces/moments

will also be positive in this direction. Forces will be taken as positive when acting in the z-direction, as will beam displacements (see Fig. 5.6).]

These equilibrium equations together with the flexibility matrix are all that we need to solve this problem. Later in this chapter, we will give a way to determine the flexibility matrix from the moment area theorem, which is a well-known beam relationship taught in a strength of materials course. For now, let us use the definition developed in Chap. 2 in terms of the elastic constants and the shape functions associated with the internal stress (see Eq. (2.28)), where

$$[G] = \int_V [N]^T [D][N] dV \tag{5.29}$$

In the beam element, we have the stress σ_{xx} given by Eq. (5.6):

(continued)

Example 5.1 (continued)

$$\sigma_{xx} = -\frac{M(\xi)z}{I} \tag{5.30}$$

where ξ is the x-distance as measured from the start of the element, whose length we will let be l. The strain $e_{xx} = \sigma_{xx}/E$, so the matrix $[D] = 1/E$, i.e., it is just a scalar constant. We can write the bending moment in the element, given by Eq. (5.17), in a matrix-vector form as

$$M(\xi) = [m_1(\xi) \;\; m_2(\xi)]\begin{Bmatrix} M_1 \\ M_2 \end{Bmatrix} = [m(\xi)]\{F\} \tag{5.31}$$

where

$$m_1(\xi) = (1 - \xi/l), \quad m_2(\xi) = \xi/l \tag{5.32}$$

and $\{F\}$ is just the column vector of generalized forces (the moments M_1 and M_2 in this case).

Thus, the stress σ_{xx} is given as

$$\sigma_{xx} = -\frac{[m(\xi)]\{F\}z}{I} = [N(\xi)]\{F\} \tag{5.33}$$

where the shape function matrix $[N(\xi)] = -[m(\xi)]z/I$. Placing this shape function into Eq. (5.29), we find

$$\begin{aligned} [G] &= \frac{1}{EI^2}\int_A z^2 dA \int_0^l [m(\xi)]^T[m(\xi)]d\xi \\ &= \frac{1}{EI}\int_0^l \begin{bmatrix} (\xi/l-1)^2 & -\xi(\xi/l-1)/l \\ -\xi(\xi/l-1)/l & \xi^2/l^2 \end{bmatrix} d\xi \\ &= \frac{l}{EI}\begin{bmatrix} 1/3 & 1/6 \\ 1/6 & 1/3 \end{bmatrix} \end{aligned} \tag{5.34}$$

where we have used the definition of the area moment I and performed the ξ integrations. Thus, for a beam, the deformation-generalized force relation $\{\Delta\} = [G]\{F\}$ is, explicitly,

(continued)

Example 5.1 (continued)

$$\begin{Bmatrix} \Delta_1 \\ \Delta_2 \end{Bmatrix} = \frac{l}{EI}\begin{bmatrix} 1/3 & 1/6 \\ 1/6 & 1/3 \end{bmatrix}\begin{Bmatrix} M_1 \\ M_2 \end{Bmatrix} \tag{5.35}$$

Later we will determine explicitly what the deformations in Eq. (5.35) represent. We now can implement the solution in MATLAB:

```
syms M1 M2 VR1 VR2 MR1 p0 L EI
Eq(1) = VR1 + (M1-M2)/L -p0*L/2;
Eq(2) = MR1 + M1 -p0*L^2/12;
Eq(3) = VR2 - (M1-M2)/L - p0*L/2;
Eq(4) = -M2 + p0*L^2/12;
T = [ M1 M2 VR1 VR2 MR1];
[E, P] = equationsToMatrix(Eq, T);
S = null(E).';
g = (L/EI)*[ 1/3 1/6; 1/6 1/3];
G = blkdiag(g, 0,0,0)     % form up the flexibility matrix and show it
 G =
 [L/(3*EI), L/(6*EI), 0, 0, 0]
 [L/(6*EI), L/(3*EI), 0, 0, 0]
 [    0,        0, 0, 0, 0]
 [    0,        0, 0, 0, 0]
 [    0,        0, 0, 0, 0]
Esys = [E; S*G];
Psys = [P; 0];
F = Esys\Psys
 F =
 -(L^2*p0)/24    % M1
 (L^2*p0)/12     % M2
 (5*L*p0)/8      % VR1
 (3*L*p0)/8      % VR2
 (L^2*p0)/8      % MR1
```

To form up the flexibility matrix, we used the MATLAB function blkdiag instead of diag, which allows us to place matrices as well as scalars along the diagonal of the flexibility matrix.

Note that this is a solution for the (M_1, M_2) parameters acting in Fig. 5.9c as well as the reactions. Thus, we have the solution to the problem in Fig. 5.7c, not the original problem in Fig. 5.7a. One way to get the solution to the original problem is to use the superposition of the two problems shown in Fig. 5.7b, c. However, since we found the reactions, we can determine the shear force and bending moment anywhere in the beam for the original

(continued)

Example 5.1 (continued)

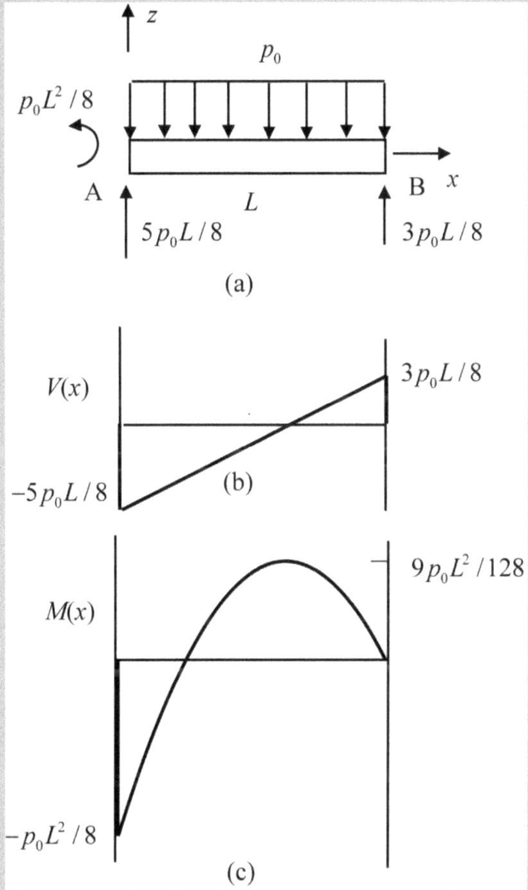

Fig. 5.10 (**a**) The beam reactions for the problem of Fig. 5.9a. (**b**) The shear force in the beam. (**c**) The bending moment in the beam

problem of Fig. 5.7a since that problem becomes statically determinate once the reactions are known. The beam with the original distributed load and the reactions are shown in Fig. 5.10a. The shear force and bending moment diagrams are shown in Fig. 5.10b, c, respectively.

5.2.3 *Distributed Forces and Their Discrete Force/Moment Equivalents*

One of the key parts in the solution process is to replace any distributed forces (and discrete forces/moments not acting at the end nodes of the elements) with equivalent forces/moments at those nodes. Figure 5.11 shows such a replacement for a distributed force. One can obtain those equivalent forces/moments by solving for a fixed-fixed beam element containing the distributed force, as done in the last example, or by using the expression obtained in Chap. 2 for equivalent forces/moments,

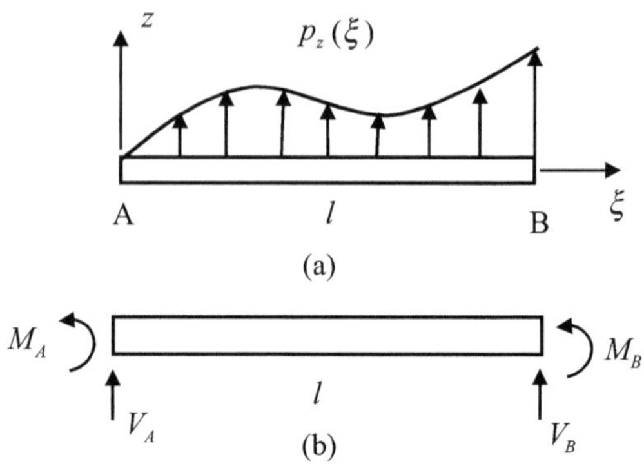

Fig. 5.11 (**a**) A distributed load and (**b**) the equivalent nodal forces/moments

Eq. (2.17), and letting the volume element $dV = Ad\xi$ and the force/unit length $p_z(\xi) = \{f\}A$, where $\{f\}$ is the body force and A is the cross-sectional area of the beam. Then we find

$$\{Q\} = \int_0^l [H]^T p_z(\xi) d\xi \tag{5.36}$$

or, more explicitly, using the displacement shape functions (see Eqs. (5.23) and (5.24)), we have (Fig. 5.11b)

$$\begin{Bmatrix} V_A \\ M_A \\ V_B \\ M_B \end{Bmatrix} = \int_0^l \begin{bmatrix} \left(1 - 3\frac{\xi^2}{l^2} + 2\frac{\xi^3}{l^3}\right) \\ \left(\xi - 2\frac{\xi^2}{l} + \frac{\xi^3}{l^2}\right) \\ \left(3\frac{\xi^2}{l^2} - 2\frac{\xi^3}{l^3}\right) \\ \left(-\frac{\xi^2}{l} + \frac{\xi^3}{l^2}\right) \end{bmatrix} p_z(\xi) d\xi \tag{5.37}$$

Figure 5.12 shows some common types of applied loads and their nodal force equivalents. Some of the expressions are rather complex, so that a MATLAB function equiv_loads was written to handle the cases shown in Fig. 5.12. The function call is:

```
[Fa, Ma, Fb, Mb] = equiv_loads(p, a, L, type);
```

where (Fa, Ma) are the equivalent force and moment acting at A and (Fb, Mb) are the corresponding force and moment at B, as shown in Fig. 5.12. For distributed forces, p is the force/unit length amplitude (acting in the negative z-direction, as shown in

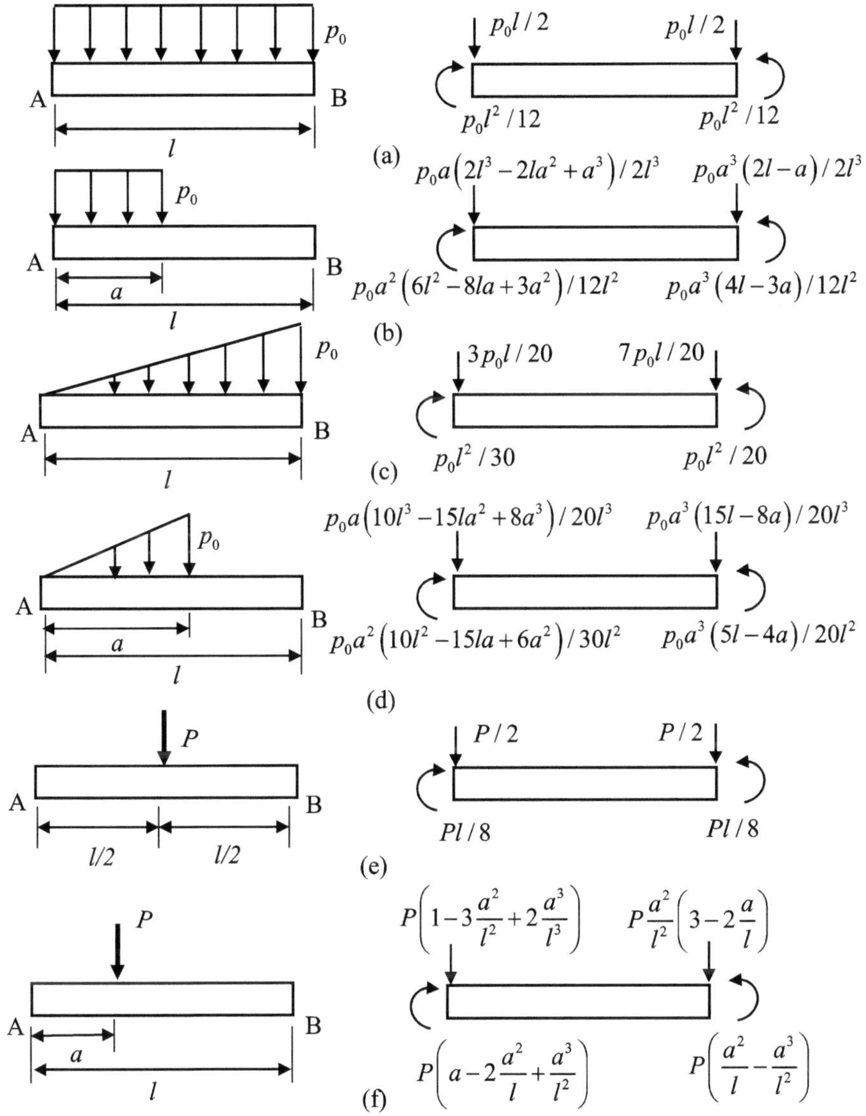

Fig. 5.12 Equivalent nodal forces/moments for: (**a**) A uniform distributed force. (**b**) A uniform distributed force over a length *a* of the element. (**c**) A linearly varying distributed force. (**d**) A linearly varying distributed force over a length *a* of the element. (**e**) A concentrated force at the center of an element. (**f**) A concentrated force at a distance *a* along the element

Fig. 5.12), a is the extent of that force, and L is the element length. For a concentrated force, p is that force and a is where the force acts in an element of length L. The argument type is a string that is either 'uniform', 'linear', or 'conc'. The listing of the MATLAB function is given in Appendix A. The arguments of the equiv_loads function can be either numerical or symbolic. Here is a symbolic example for a concentrated force P located at a distance $a = L/4$ on an element of length L:

```
syms P L
[Fa, Ma, Fb, Mb] = equiv_loads(P, L/4, L, 'conc')
Fa = (27*P)/32
Ma = (9*L*P)/64
Fb = (5*P)/32
Mb = (3*L*P)/64
```

Treating the applied distributed force/unit length $p_z(\xi)$ as a special case of a body force distribution was the approach used in Chap. 4 for axial force and torsion problems and was followed in a similar manner here to obtain Eq. (5.37). However, one might object to that approach since the forces applied to beams are usually applied to the surface of the beam (often the top surface) and are not distributed throughout the volume, as gravity is, for example. Thus, physically, it makes more sense to consider the distributed load as arising from the surface stress distributions and write the nodal forces as (see Eq. 2.7):

$$\{P_S\} = \int_S [H]^T \left\{T^{(\mathbf{n})}\right\} dS \tag{5.38}$$

One can treat this surface integral as a line integral around the periphery C of the cross-sectional area A of the beam and an integral in the ξ-direction as

$$\{P_S\} = \int_0^l [H(\xi)]^T \left[\int_C \left\{T^{(\mathbf{n})}\right\} dC \right] d\xi \tag{5.39}$$

and define the force/unit length as

$$p_z(\xi) = \int_C \left\{T^{(\mathbf{n})}\right\} dC \tag{5.40}$$

which leads us to the same expression found in Eq. (5.37).

5.2.4 The Force-Based Method for Beams

The force-based method for beams follows very similar steps to those found for axial forces and torques:

1. The beam is broken up into elements.
2. The applied forces are replaced by equivalent nodal forces/moments at the ends of the elements.

3. All the forces/moments at the nodes are shown in free body diagrams for the nodes. These include the equivalent forces/moments, any directly applied nodal forces/moments, and the internal shear force and bending moment acting on the node from adjacent elements. In beam problems, we have used the internal moment variables (M_{2n-1}, M_{2n}) to describe the linear varying internal moment distribution for the nth element. The constant shear force in the nth element is given by $V_n = (M_{2n-1} - M_{2n})/l_n$, where l_n is the length of the nth element.
4. Equilibrium equations are written symbolically in MATLAB for all the nodes. The vector $\{F\}$ is identified as the symbolic unknowns of the problem. Those unknowns can include internal forces/moments and reaction forces/moments. MATLAB is used to convert these symbolic equations to the equilibrium matrix $[E]$ and load vector $\{P\}$ in the equilibrium eqs. $[E]\{F\} = \{P\}$.
5. Homogeneous solutions of the equilibrium equations are found with the MATLAB null function and used to obtain the compatibility matrix $[S]$.
6. The compatibility matrix $[G]$ is formed from the flexibilities in the problem. For beams the flexibility matrix for the internal moment variables in the nth element is

$$\left[G^{(n)}\right] = \frac{l_n}{EI}\begin{bmatrix} 1/3 & 1/6 \\ 1/6 & 1/3 \end{bmatrix}$$

   where Young's modulus, E, and the area moment I, can also be different for each element, in general. The flexibilities of reaction forces/moments are just zeros. The MATLAB function blkdiag can be used to generate the complete flexibility matrix for all the elements.
7. The equilibrium matrix, load vector, compatibility matrix, and flexibility matrix are used to form the system equations, which can be solved for the forces $\{F\}$:

$$\begin{bmatrix} [E] \\ - \\ [S][G] \end{bmatrix}\{F\} = \begin{Bmatrix} \{P\} \\ - \\ \{0\} \end{Bmatrix}$$

   The number of zeros in $\{0\}$ is equal to the number of compatibility equations (number of rows in $[S]$).
8. After the forces are obtained the displacements, $\{U\}$, if needed, can be obtained from

$$\left[E_{sys}\right]^T \begin{Bmatrix} \{-U\} \\ - \\ \{0\} \end{Bmatrix} = [G]\{F\}$$

This solution process again is very similar to the method of pins for trusses and the nodal method used for axial forces and torsion problems. The only major difference is that the element flexibility is defined in terms of a 2 × 2 matrix rather than just a scalar.

5.2.5 A Force-Based Solution and Singularity Functions

Once the unknown beam reaction forces from supports have been found with the force-based method the problem is statically determinate and we can find the internal shear force and bending moment anywhere in the beam. Even the displacements and slopes everywhere in the beam can be found. As with axial force and torsion problems, singularity functions make these distributions easy to obtain and plot. Figure 5.13 shows some types of force distributions and the associated singularity functions that define the internal shear force and bending moment. The insert in Fig. 5.13 gives the directions assumed for the internal shear force and bending moment.

to remind you of those assumptions. Some texts change those assumed directions from the ones given here, so it is wise to check the assumptions being made on the shear force and bending moment directions.

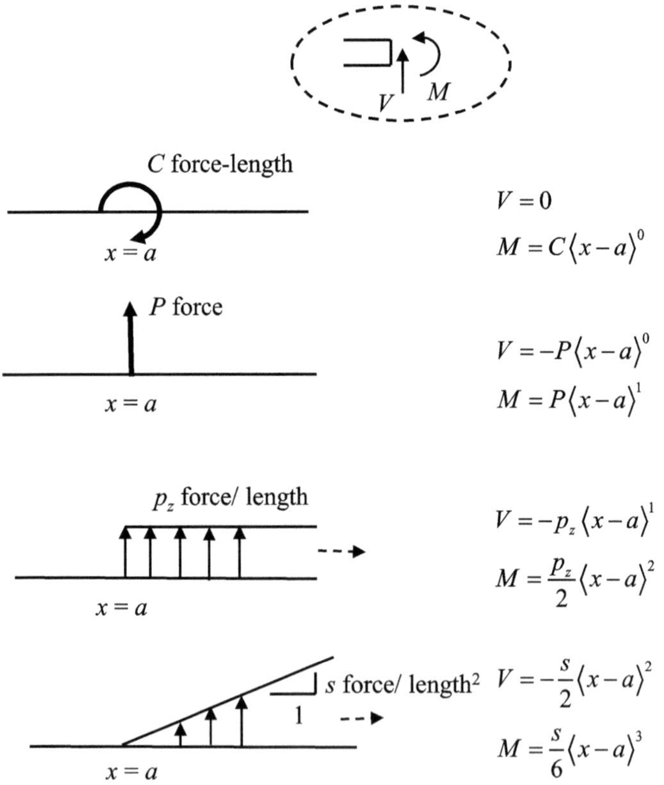

Fig. 5.13 Some applied forces and couples and distributed forces acting on a beam and the singularity functions associated with those applied loads that describe the internal shear force and bending moment. As a reminder, the assumed directions of the internal shear force and bending moment are shown in the dashed insert

Example 5.2: A Fixed-Fixed Beam and Singularity Functions

Let's now solve a problem and plot the results with singularity functions. Consider the beam shown in Fig. 5.14a. Let p_0= 20 lb./ft., L = 3 ft., and assume EI is constant in the beam. Solve this statically indeterminate problem and plot the internal shear force and bending moment with singularity functions. The internal forces/moments are shown in Fig. 5.14b and Fig. 5.14c shows the free body diagrams of the three nodes, including the equivalent forces/moments from the distributed force. From those free body diagrams, we have the symbolic equilibrium equations in MATLAB given by.

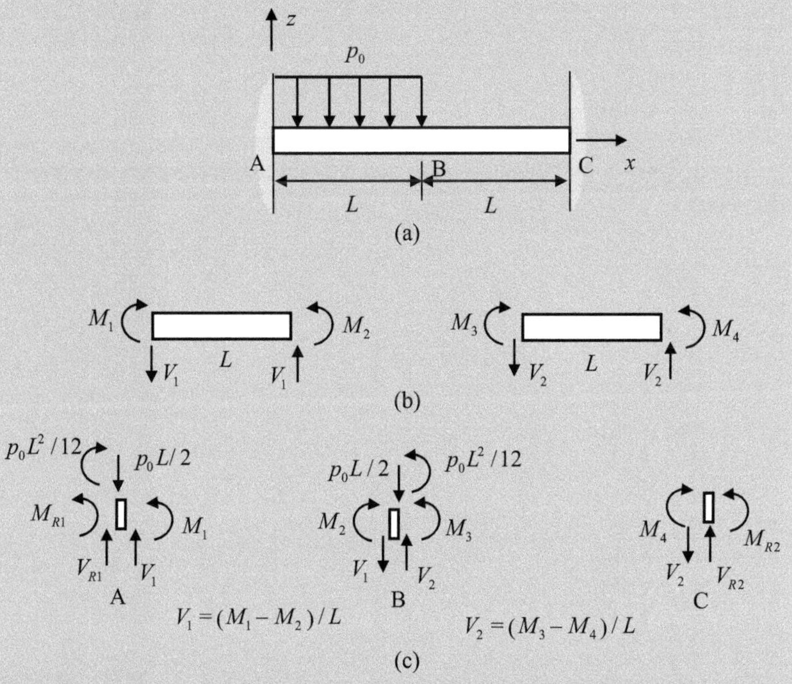

Fig. 5.14 (**a**) A statically indeterminate problem. (**b**) Two beam elements. (**c**) The free body diagrams of the nodes

```
syms M1 M2 M3 M4 VR1 MR1 VR2 MR2
T= [M1 M2 M3 M4 VR1 MR1 VR2 MR2];
Eq(1) =VR1+(M1 - M2)/3 -30 == 0;
Eq( 2) =MR1+M1 - 15 == 0;
Eq(3) = -(M1 - M2)/3 +(M3 - M4)/3 -30 == 0;
Eq(4) =M3 - M2 + 15 == 0;
Eq(5) = -(M3 - M4)/3 + VR2 == 0;
Eq(6) =MR2 - M4 == 0;
```

(continued)

Example 5.2 (continued)

Now, convert to the equilibrium matrix and force vector and solve for the compatibility matrix:

```
[E, P] = equationsToMatrix(Eq, T);
S = null(E).';
```

Next, form the flexibility matrix and the system equations:

```
g1 = [1/3 1/6; 1/6 1/3];        % omit the common L/EI term
g2 = g1;
G = blkdiag(g1, g2, 0, 0, 0, 0);
Esys = [E; S*G];
Psys = [P; 0 ; 0];
```

Finally, solve for all the unknowns and convert to double-precision numerical values:

```
F = Esys\Psys;
double(F)
ans =

 -26.2500  % M1
  30.0000  % M2
  15.0000  % M3
-18.7500   %M4
 48.7500   % VR1
 41.2500   % MR1
 11.2500   % VR2
-18.7500   % MR2
```

The reactions at the fixed supports are shown in Fig. 5.15a. If we calculate and plot the shear force with singularity functions, we have.

```
V = @(x) -48.75*step_sf(x, 0) + 20*lin_sf(x, 0) - 20*lin_sf(x, 3)
-11.25*step_sf(x, 6);
x = linspace(0, 6, 1000);
plot(x, V(x))
xlabel('x, ft')
ylabel(' V, lb')
```

(continued)

Example 5.2 (continued)

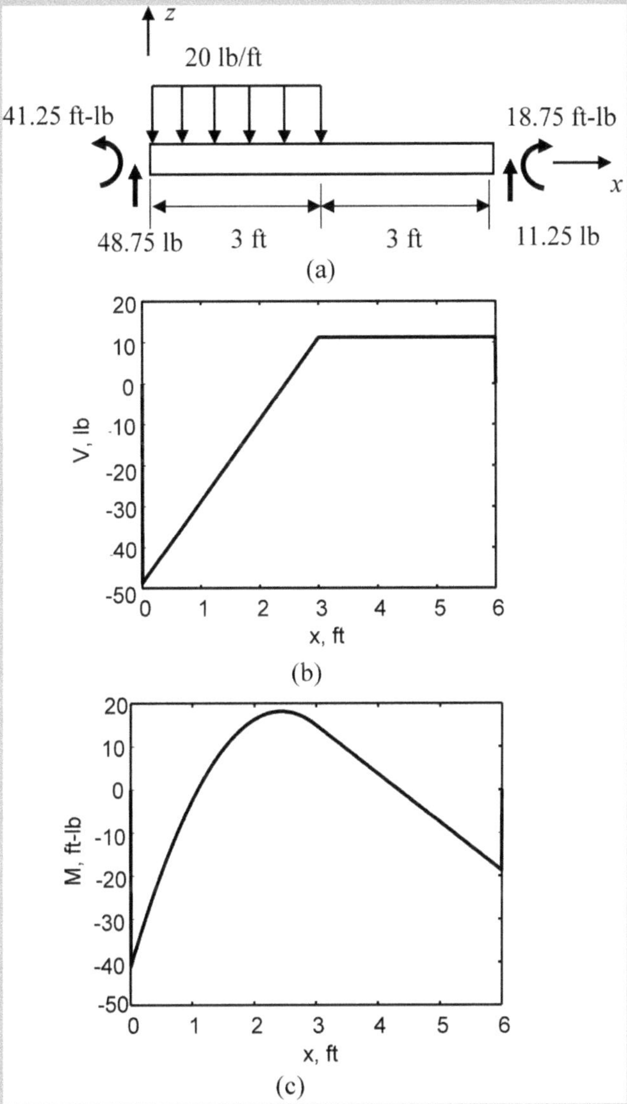

Fig. 5.15 (**a**) The forces acting on the fixed-fixed beam. (**b**) The shear force *V*(*x*). (**c**) The bending moment *M*(*x*)

The plot is shown in Fig. 5.15b. Similarly, for the bending moment,

```
M = @(x) -41.25*step_sf(x,0) + 48.75*lin_sf(x, 0) -10*quad_sf(x,
0) ...
+ 10*quad_sf(x, 3) + 11.25*lin_sf(x, 6) + 18.75*step_sf(x, 6);
```

(continued)

Example 5.2 (continued)

```
plot(x, M(x))
xlabel(' x, ft')
ylabel(' M, ft-lb')
```

which is plotted in Fig. 5.15c. Since the bending moment is the negative of the integral of the shear force, we could obtain it by integrating the negative of the shear force terms and then adding in any concentrated couple singularity functions present. The shear force and the bending moment both start at zero and end at zero. Note that the term 11.25*lin_sf(x, 6) in the moment expression could be omitted since it is zero for $0 \le x \le 6$ ft. The anonymous functions given here can also be used to obtain values for the shear force and bending moment at any specific points in the beam.

5.2.6 Deformation-Displacement Relations for Beams

Although we can solve for the force/moments in a beam without directly involving the displacements, the deformation-displacement relation is still a part of the solution in that it is the reason why we can obtain the compatibility equations from homogeneous solutions of the equilibrium equations. Thus, in this section, we want to examine the deformation-displacement relation in more detail. As a part of that examination, we will see that we can also obtain a physical interpretation for the deformations that appear in beams as part of the algebraic equations of linear elasticity. We will use the problem of Fig. 5.9a as the focus of our discussion since we solved that problem with a single element so we can focus on the deformation-displacement relations for any element in the same fashion. We found the equilibrium equations for the beam of Fig. 5.7a in Eq. (5.28). If that set of equations is placed in matrix-vector form, we have

$$\begin{bmatrix} 1/L & -1/L & 1 & 0 & 0 \\ 1 & 0 & 0 & 1 & 0 \\ -1/L & 1/L & 0 & 0 & 1 \\ 0 & -1 & 0 & 0 & 0 \end{bmatrix} \begin{Bmatrix} M_1 \\ M_2 \\ V_{R1} \\ M_{R1} \\ V_{R2} \end{Bmatrix} = \begin{Bmatrix} p_0 L/2 \\ p_0 L^2/12 \\ p_0 L/2 \\ -p_0 L^2/12 \end{Bmatrix} \tag{5.41}$$

We want to focus first on the equilibrium equations for the internal forces, so let's place all the unknown reactions on the right side of the equilibrium equations, to obtain

$$\begin{bmatrix} 1/L & -1/L \\ 1 & 0 \\ -1/L & 1/L \\ 0 & -1 \end{bmatrix} \begin{Bmatrix} M_1 \\ M_2 \end{Bmatrix} = \begin{Bmatrix} p_0 L/2 - V_{R1} \\ p_0 L^2/12 - M_{R1} \\ p_0 L/2 - V_{R2} \\ -p_0 L^2/12 \end{Bmatrix} \tag{5.42}$$

Thus, for our single element, if we write these equations as $[E]\{F\} = \{P\}$, where $\{F\} = [M_1 \quad M_2]^T$, then the equilibrium matrix for the element is

$$[E] = \begin{bmatrix} 1/L & -1/L \\ 1 & 0 \\ -1/L & 1/L \\ 0 & -1 \end{bmatrix} \tag{5.43}$$

In the algebraic equations of linear elasticity, the deformation-displacement relations are written as $\{\Delta\} = [E]^T\{U\}$, so for our beam element, we have

$$\begin{Bmatrix} \Delta_1 \\ \Delta_2 \end{Bmatrix} = \begin{bmatrix} 1/L & 1 & -1/L & 0 \\ -1/L & 0 & 1/L & -1 \end{bmatrix} \begin{Bmatrix} -w_1 \\ -\theta_1 \\ -w_2 \\ -\theta_2 \end{Bmatrix} \tag{5.44}$$

where the minus signs occur because of our choice of how we found the equations of equilibrium, as discussed previously. Thus, we have two deformations associated with the moments (M_1, M_2) given by

$$\begin{aligned} \Delta_1 &= w_2/L - w_1/L - \theta_1 \\ \Delta_2 &= w_1/L - w_2/L + \theta_2 \end{aligned} \tag{5.45}$$

To give a physical interpretation of these deformations, we need to discuss the *second moment area theorem* for beams (do not confuse the term "second moment area" with second "area moment"). Consider Fig. 5.16a, which shows a segment of a beam of length L. Also shown are the tangents and normal to the neutral axis at A and B, where angle θ is common the angle between both those two tangents and the corresponding two normals. We will let $t_{B/A}$ be the vertical distance measured at B between the two tangents at A and B, as shown in Fig. 5.16a. Now consider a small element of the beam at a location x from A (Fig. 5.16b). Since the slopes of the beam are assumed to be everywhere small [Note: the slopes are not shown as small in Figs. 5.16, 5.17, 5.18, and 5.19 only so they can be more easily displayed], we have $dt_{B/A} = (L - x)d\theta$. Also, we can write the moment-curvature relationship as

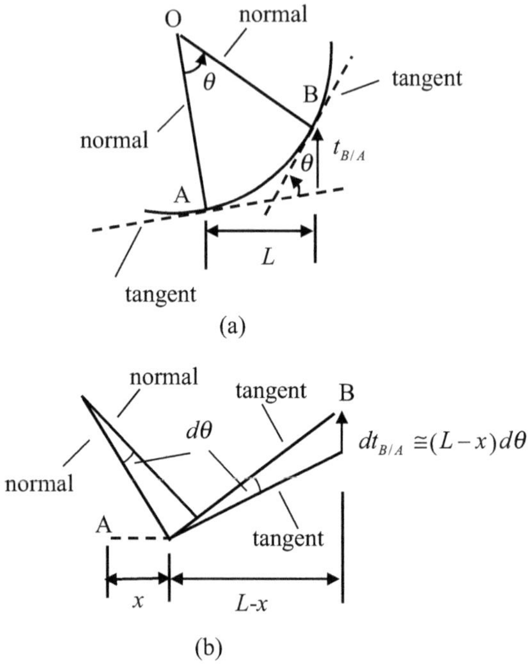

Fig. 5.16 (**a**) The geometry of a beam, showing the vertical distance at B between tangent lines at A and B. (**b**) Contribution to the vertical distance shown in (**a**) for a small segment at a distance *x* from A

$$\frac{d^2w}{dx^2} = \frac{d\theta}{dx} = \frac{M(x)}{EI} \tag{5.46}$$

since $\theta = dw/dx$ for small slopes.

From Fig. 5.16b and Eq. (5.46), we find

$$\frac{d\theta}{dx} = \frac{M(x)}{EI}, \quad dt_{B/A} = (L-x)d\theta \tag{5.47}$$

so that we have

$$t_{B/A} = \int_A^B dt_{B/A} = \frac{1}{EI}\int_0^L (L-x)M(x)dx \tag{5.48}$$

We can examine the distance between tangents as measured at A in the same fashion, as shown in Fig. 5.17. In this case, we find

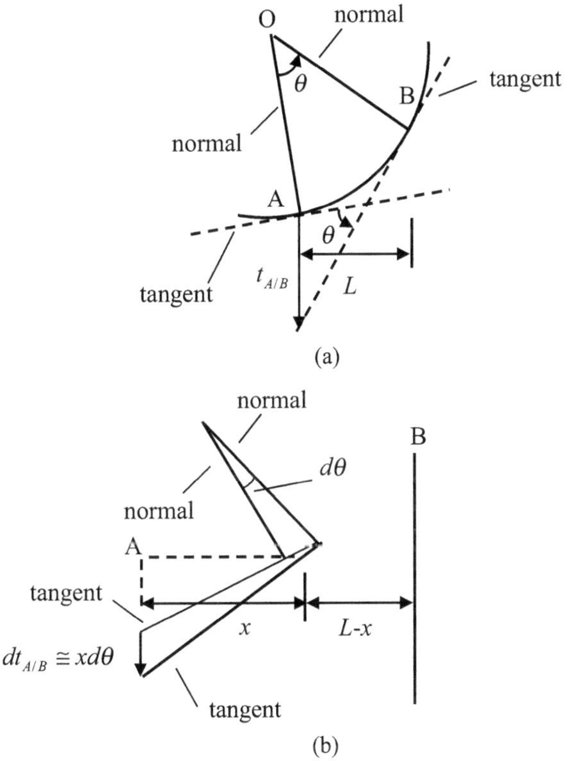

Fig. 5.17 (**a**) The geometry of a beam, showing the vertical distance at A between tangent lines at A and B. (**b**) Contribution to the vertical distance shown in (**a**) for a small segment at a distance x from A

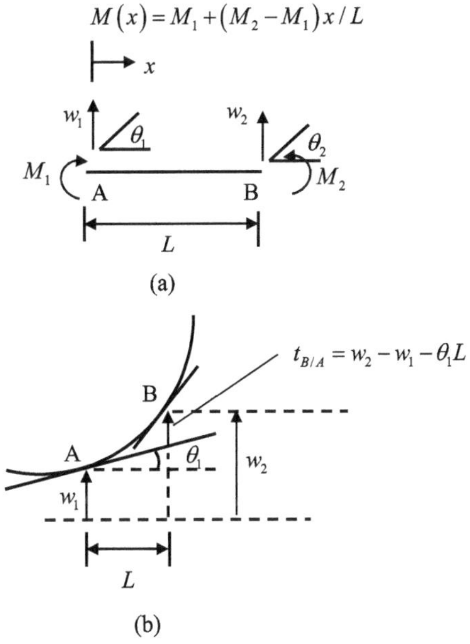

Fig. 5.18 (**a**) A beam element, showing the moment parameters and nodal displacements. (**b**) The distance between tangents at B in terms of the nodal displacements

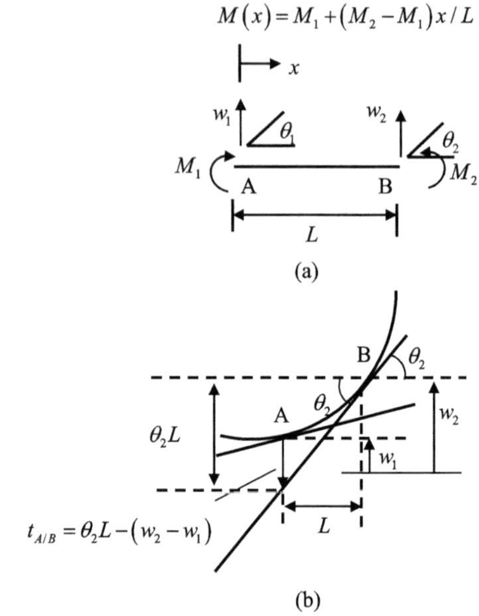

Fig. 5.19 (**a**) A beam element, showing the moment parameters and nodal displacements. (**b**) The distance between tangents at A in terms of the nodal displacements

$$t_{A/B} = \int dt_{A/B} = \frac{1}{EI}\int_{0}^{L} xM(x)dx \tag{5.49}$$

Now, let's connect these results to Eq. (5.45). Figure 5.18a shows an element of a beam where the bending moment is defined in terms of (M_1, M_2). As can be seen from the geometry in Fig. 5.18b, we have

$$t_{B/A} = (w_2 - w_1) - \theta_1 L \tag{5.50}$$

Similarly, from Fig. 5.19b, we find

$$t_{A/B} = \theta_2 L - (w_2 - w_1) \tag{5.51}$$

Comparing these results with Eq. (5.45), we find

$$\begin{aligned}\Delta_1 &= t_{B/A}/L = w_2/L - w_1/L - \theta_1\\ \Delta_2 &= t_{A/B}/L = w_1/L - w_2/L + \theta_2\end{aligned} \tag{5.52}$$

so we have found the physical meaning of the deformations in terms of these distances between tangents in an element. Furthermore, using Eqs. (5.48) and (5.49), we obtain

$$\begin{aligned} t_{B/A} &= \frac{1}{EI}\int_0^L (L-x)M(x)dx = \frac{L^2}{EI}(M_1/3 + M_2/6) \\ t_{A/B} &= \frac{1}{EI}\int_0^L xM(x)dx = \frac{L^2}{EI}(M_1/6 + M_2/3) \end{aligned} \tag{5.53}$$

which gives

$$\begin{Bmatrix} \Delta_1 \\ \Delta_2 \end{Bmatrix} = \frac{L}{EI}\begin{bmatrix} 1/3 & 1/6 \\ 1/6 & 1/3 \end{bmatrix}\begin{Bmatrix} M_1 \\ M_2 \end{Bmatrix} \tag{5.54}$$

The expressions for the deformations as the normalized distances between tangents in Eq. (5.52), therefore also show that these deformations are related to the moment parameters (M_1, M_2) in terms of the same flexibility matrix defined earlier.

Now, let's go back to the original equilibrium matrix for the problem of Fig. 5.7a where the reactions were included in the equilibrium matrix (see Eq. (5.41)). The deformation-displacement relation for this case, therefore, is

$$\begin{Bmatrix} \Delta_1 \\ \Delta_2 \\ \Delta_{VR1} \\ \Delta_{MR1} \\ \Delta_{VR2} \end{Bmatrix} = \begin{bmatrix} 1/L & 1 & -1/L & 0 \\ -1/L & 0 & 1/L & -1 \\ 1 & 0 & 0 & 0 \\ 0 & 1 & 0 & 0 \\ 0 & 0 & 1 & 0 \end{bmatrix}\begin{Bmatrix} -w_1 \\ -\theta_1 \\ -w_2 \\ -\theta_2 \end{Bmatrix} \tag{5.55}$$

The deformations (Δ_1, Δ_2) are again given by Eq. (5.52) but since we must set the deformations of the reactions $\Delta_{VR1} = \Delta_{MR1} = \Delta_{VR2} = 0$, we also have from Eq. (5.55) $w_1 = \theta_1 = w_2 = 0$, so we have also satisfied the displacement boundary conditions at the supports, leaving the only free nodal displacement θ_2.

5.3 Solving for Displacements in Beams and the Pseudo Stiffness Matrix

Example 5.3: A Statically Indeterminate Beam-Solving for Displacements

If the forces are known, we can solve for the displacements, as shown previously (see Eq. (1.29)). Let's use the problem of Fig. 5.7a as an example. Recall the equilibrium equations for this problem were given by Eq. (5.41):

$$\begin{bmatrix} 1/L & -1/L & 1 & 0 & 0 \\ 1 & 0 & 0 & 1 & 0 \\ -1/L & 1/L & 0 & 0 & 1 \\ 0 & -1 & 0 & 0 & 0 \end{bmatrix} \begin{Bmatrix} M_1 \\ M_2 \\ V_{R1} \\ M_{R1} \\ V_{R2} \end{Bmatrix} = \begin{Bmatrix} p_0L/2 \\ p_0L^2/12 \\ p_0L/2 \\ -p_0L^2/12 \end{Bmatrix} \tag{5.56}$$

The stiffness matrix is

$$[G] = \frac{L}{EI}\begin{bmatrix} 1/3 & 1/6 & 0 & 0 & 0 \\ 1/6 & 1/3 & 0 & 0 & 0 \\ 0 & 0 & 0 & 0 & 0 \\ 0 & 0 & 0 & 0 & 0 \\ 0 & 0 & 0 & 0 & 0 \end{bmatrix} \tag{5.57}$$

Now, we can solve the problem for the forces and displacements in MATLAB:

```
syms p L EI
E = [ 1/L -1/L 1 0 0; 1 0 0 1 0; -1/L 1/L 0 0 1;...
0 -1 0 0 0];
P = [ p*L/2; p*L^2/12; p*L/2; -p*L^2/12];
g = (L/EI)*[1/3 1/6 ; 1/6 1/3];
G = blkdiag(g, 0, 0, 0);
S = null(E).';
Esys = [ E; S*G];
Psys = [ P;0];
```

(continued)

Example 5.3 (continued)

```
F = Esys\Psys
F =

-(L^2*p)/24   % M1
 (L^2*p)/12   % M2
 (5*L*p)/8    % VR1
 (L^2*p)/8    % MR1
 (3*L*p)/8    % VR2
```

This agrees with our previous results (although the order of the reactions is different). Now, let's find the displacements (there is only one nonzero displacement—the rotation at the right end):

```
U = Esys.'\G*F
U =

 0                  % - U1 at A
 0                  % - theta1 at A
 0                  % - U2 at B
-(L^3*p)/(48*EI)    % - theta2 at B
 0                  % additional zero
```

so the rotation at the right support is $\theta_2 = p_0 L^3/48EI$.

If we use the pseudo stiffness matrix instead to find this displacement, we must modify the flexibility matrix so that it does not contain zeros. We must also keep the common L/EI factor in all terms to maintain the consistency of the units:

```
Gm = blkdiag(g, L*10^-6/EI, L*10^-6/EI, L*10^-6/EI);
K = E*inv(Gm)*E.';
Um = K\P    % solve for approximate displacements
Um =

 (L^2*p*(1250003*L^2 + 6))/(2000000*EI*(1000003*L^2 + 6))
              (L^5*p)/(8*EI*(1000003*L^2 + 6))
(3*L^2*p*(250001*L^2 + 2))/(2000000*EI*(1000003*L^2 + 6))
   -(L^3*p*(500003*L^2 + 12))/(24*EI*(1000003*L^2 + 6))
```

Because this result is symbolic, it is difficult to read. But note that all the terms except the last are very small, so neglecting all the insignificant terms and digits, we have approximately.

(continued)

Example 5.3 (continued)

```
Um =
0                  % -U1 at A
0                  % -theta1 at A
0                  % -U2 at B
-5pL^3/24*EI*10    % -theta2 at B
```

which agrees with our previous result, and there is no extra zero.

5.4 The Force-Based Method for Frames

Two-dimensional frames contain combinations of bending and axial forces and three-dimensional frames may involve extensions, bending, and torsion, so they can involve many unknowns. Despite this complexity, we can still analyze simple frames by examining all the forces at the nodes between members, just as done in trusses.

Example 5.4: An L-Shaped Frame

Consider, for example, the pinned L-shaped frame in Fig. 5.20a. We want to determine the reactions at the pins. Let the bending flexibility be $g_b = L/EI$ and the extensional flexibility be $g_a = L/EA$.

Taking two elements of length L, the force/moments acting at nodes A, B, and C are shown in Fig. 5.20b. Here are the equations of equilibrium at the nodes and the MATLAB solution:

$$
\begin{aligned}
{}_A\sum F_x &= 0 \quad A_x - (M_1 - M_2)/L = 0 \\
{}_A\sum F_y &= 0 \quad A_y + N_1 = 0 \\
{}_A\sum M_z &= 0 \quad M_1 = 0 \\
{}_C\sum F_x &= 0 \quad N_2 + (M_1 - M_2)/L = 0 \\
{}_C\sum F_y &= 0 \quad -N_1 - P/2 + (M_3 - M_4)/L = 0 \\
{}_C\sum M_z &= 0 \quad M_3 - M_2 - PL/8 = 0 \\
{}_B\sum F_x &= 0 \quad B_x - N_2 = 0 \\
{}_B\sum F_y &= 0 \quad B_y - (M_3 - M_4)/L - P/2 = 0 \\
{}_B\sum M_z &= 0 \quad -M_4 + PL/8 = 0
\end{aligned}
$$

(continued)

Example 5.4 (continued)

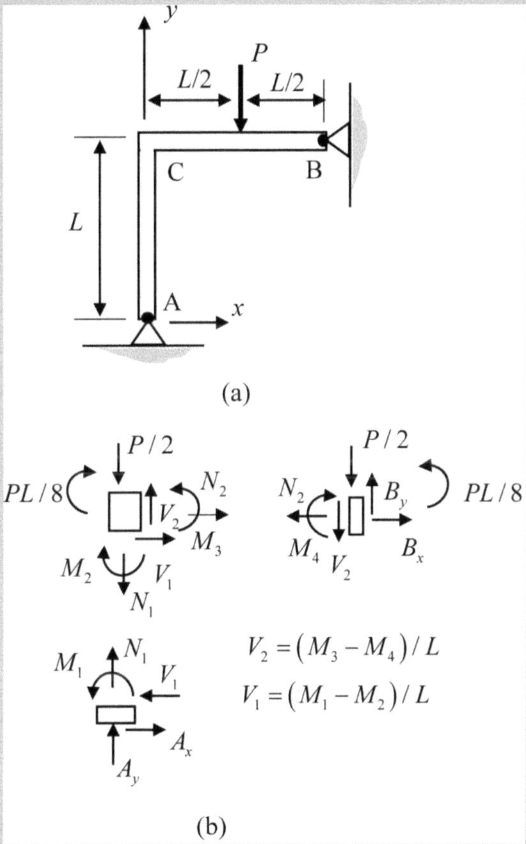

Fig. 5.20 (**a**) An L-shaped frame. (**b**) The forces/moments acting at the nodes

```
syms M1 M2 M3 M4 N1 N2 Ax Ay Bx By P L
Eq(1) = Ax - (M1-M2)/L;
Eq(2) = Ay + N1;
Eq(3) = M1;
Eq(4) = (M1 - M2)/L + N2;
Eq(5) = (M3 - M4)/L - N1 - P/2;
Eq(6) = M3 - M2 - P*L/8;
Eq(7) = Bx - N2;
Eq(8) = By - (M3 - M4)/L - P/2;
Eq(9) = - M4 + P*L/8;
T = [ M1 M2 M3 M4 N1 N2 Ax Ay Bx By];
[E, P] = equationsToMatrix(Eq, T);
S = null(E).';
syms ga gb
flexb = gb*[1/3 1/6; 1/6 1/3];
G = blkdiag(flexb, flexb, ga, ga, 0, 0, 0, 0);
Esys = [ E; S*G];
```

(continued)

Example 5.4 (continued)

```
Psys = [P; 0];
F = Esys\Psys
F =
 0                                             % M1
(3*L*(- P*gb*L^2 + 8*P*ga))/(32*(gb*L^2 + 3*ga)) % M2
 (L*(P*gb*L^2 + 36*P*ga))/(32*(gb*L^2 + 3*ga))  % M3
 (L*P)/8                                        % M4
 -(19*P*gb*L^2 + 24*P*ga)/(32*(gb*L^2 + 3*ga))  % N1
 (3*(- P*gb*L^2 + 8*P*ga))/(32*(gb*L^2 + 3*ga)) % N2
 -(3*(- P*gb*L^2 + 8*P*ga))/(32*(gb*L^2 + 3*ga)) % Ax
 (19*P*gb*L^2 + 24*P*ga)/(32*(gb*L^2 + 3*ga))   % Ay
 (3*(- P*gb*L^2 + 8*P*ga))/(32*(gb*L^2 + 3*ga)) % Bx
 (13*P*gb*L^2 + 72*P*ga)/(32*(gb*L^2 + 3*ga))   % By
```

This answer in symbolic form is rather complex. However, in many frames, the extensional flexibility is much smaller than the bending flexibility, so if we set the extensional flexibility to be zero, we obtain results for the bending contributions only, and those are independent of the bending flexibility. Here are the frame reactions in that case:

```
subs( F(7), ga, 0)
 ans = (3*P)/32        % Ax
 subs( F(8), ga, 0)
 ans = (19*P)/32       % Ay
 subs( F(9), ga, 0)
 ans = -(3*P)/32       % Bx
 subs( F(10), ga, 0)
 ans = (13*P)/32       % By
```

The reactions are shown in Fig. 5.21. They satisfy overall force and moment equilibrium for the frame. With these reactions known, it is possible to obtain the internal forces but we will not show those results here.

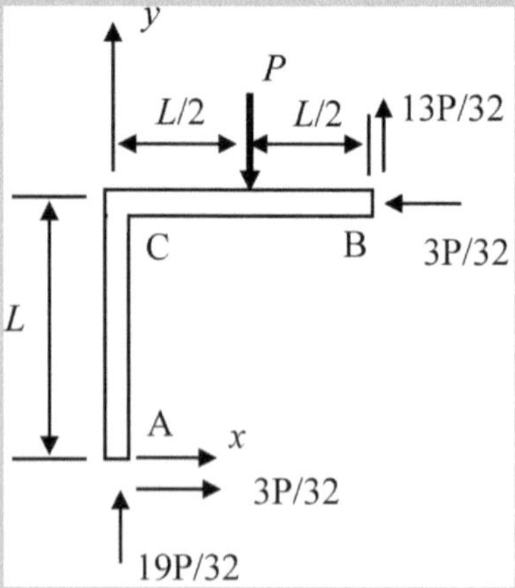

Fig. 5.21 The reactions at the pins of the L-frame of Fig. 5.20a

Example 5.5: A V-Frame

Consider now the V-frame shown in Fig. 5.22a. It is pinned at both ends and carries a uniform distributed force/unit length p_0 acting normal to side AC as shown. Unlike a truss, the applied forces are not acting at nodes and the sides are now connected rigidly at C so that the members are both in a combination of extension and bending, as in the last problem. We want to determine the reactions at the pins. Let the bending flexibility be $g_b = L/EI$ and the extensional flexibility be $g_a = L/EA$. In this case, we will set the extensional flexibility to be zero at the outset so that only bending effects are present. Taking the sides of the frame as two elements of length L, the forces/moments at the nodes are as shown in Fig. 5.22b. Here are the equilibrium equations at the nodes and the MATLAB solution:

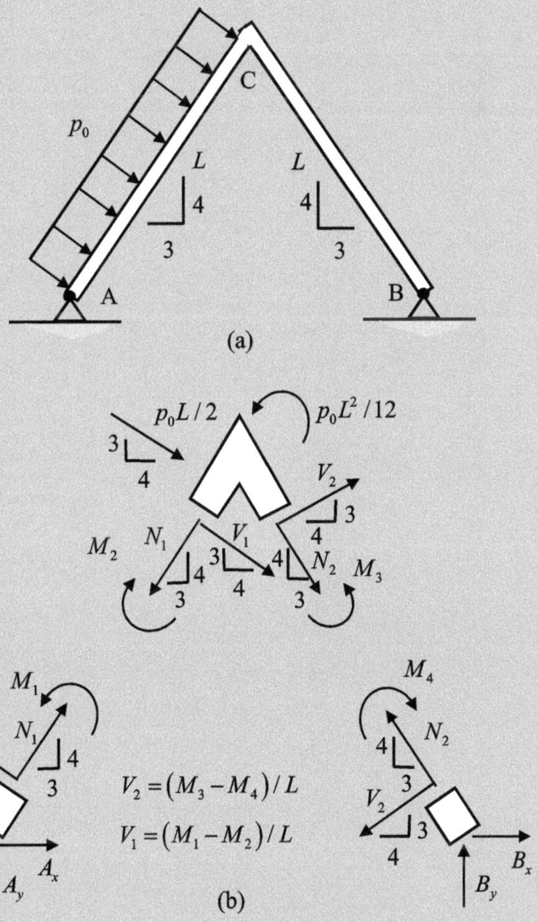

Fig. 5.22 (**a**) A V-frame. (**b**) The forces/moments acting at the nodes

(continued)

Example 5.5 (continued)

$$
\begin{aligned}
{}_{A}\sum F_x=0 &\quad A_x+3N_1/5-(4/5)(M_1-M_2)/L+(4/5)(p_0L/2)=0\\
{}_{A}\sum F_y=0 &\quad A_y+4N_1/5+(3/5)(M_1-M_2)/L-(3/5)(p_0L/2)=0\\
{}_{A}\sum M_z=0 &\quad M_1-p_0L^2/12=0\\
{}_{C}\sum F_x=0 &\quad -3N_1/5+(4/5)(M_1-M_2)/L+3N_2/5\\
&\quad +(4/5)(M_3-M_4)/L+(4/5)(p_0L/2)=0\\
{}_{C}\sum F_y=0 &\quad -4N_1/5-(3/5)(M_1-M_2)/L-4N_2/5\\
&\quad +(3/5)(M_3-M_4)/L-(3/5)(p_0L/2)=0\\
{}_{C}\sum M_z=0 &\quad M_3-M_2+p_0L^2/12=0\\
{}_{B}\sum F_x=0 &\quad B_x-3N_2/5-(4/5)(M_3-M_4)/L=0\\
{}_{B}\sum F_y=0 &\quad B_y+4N_2/5-(3/5)(M_3-M_4)/L=0\\
{}_{B}\sum M_z=0 &\quad -M_4=0
\end{aligned}
$$

```
syms M1 M2 M3 M4 N1 N2 Ax Ay Bx By p0 L gb
Eq(1) = Ax + 3*N1/5 - (4/5)*(M1 - M2)/L + (4/10)*p0*L;
Eq(2) = Ay + 4*N1/5 + (3/5)*(M1 - M2)/L - (3/10)*p0*L;
Eq(3) = M1 - p0*L^2/12;
Eq(4) = -(3/5)*N1 + (4/5)*(M1- M2)/L + (3/5)*N2 ...
 + (4/5)*(M3 - M4)/L + (4/10)*p0*L;
Eq(5) = -(4/5)*N1 - (3/5)*(M1 - M2)/L - (4/5)*N2 ...
 + (3/5)*(M3 - M4)/L - (3/10)*p0*L;
Eq(6) = M3 - M2 + p0*L^2/12;
Eq(7) = Bx - (3/5)*N2 - (4/5)*(M3 - M4)/L;
Eq(8) = (4/5)*N2 - (3/5)*(M3 - M4)/L + By;
Eq(9) = M4;
T = [ M1 M2 M3 M4 N1 N2 Ax Ay Bx By];
[E, P] = equationsToMatrix(Eq, T);
S = null(E).';
fb = gb*[1/3 1/6; 1/6 1/3];
G = blkdiag(fb, fb, 0, 0, 0, 0, 0, 0);
Esys = [E; S*G];
Psys = [P;0];
F = Esys\Psys
F =
   (L^2*p0)/12    % M1
```

(continued)

Example 5.5 (continued)

```
(L^2*p0)/48        % M2
-(L^2*p0)/16       % M3
     0             % M4
(19*L*p0)/192      % N1
-(109*L*p0)/192    % N2
-(131*L*p0)/320    % Ax
(11*L*p0)/60       % Ay
-(25*L*p0)/64      % Bx
 (5*L*p0)/12       % By
```

The reactions found are shown in Fig. 5.23. One can verify that they satisfy force and moment equilibrium for the entire frame.

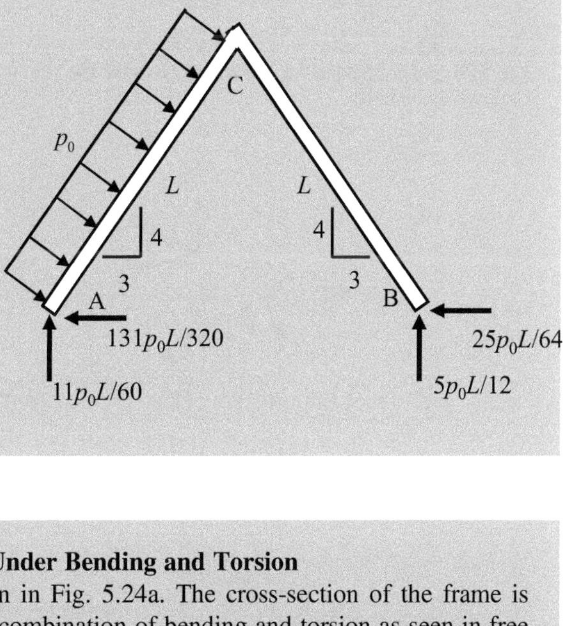

Fig. 5.23 The reactions at the pins of the V-frame

Example 5.6: A Frame Under Bending and Torsion

Consider the frame shown in Fig. 5.24a. The cross-section of the frame is circular, and it is under a combination of bending and torsion as seen in free body diagrams of the two elements AB and BC in Fig. 5.24b. The length $L = 4$ ft. and let $GJ = 2EI/3$, where G is the shear modulus, J is the polar area moment, E is Young's modulus, and I is the area moment about a horizontal axis through the centroid of the cross-section. The forces and moments acting at the nodes A, B, and C are shown in Fig. 5.25. From equilibrium at those nodes, we have

(continued)

Example 5.6 (continued)

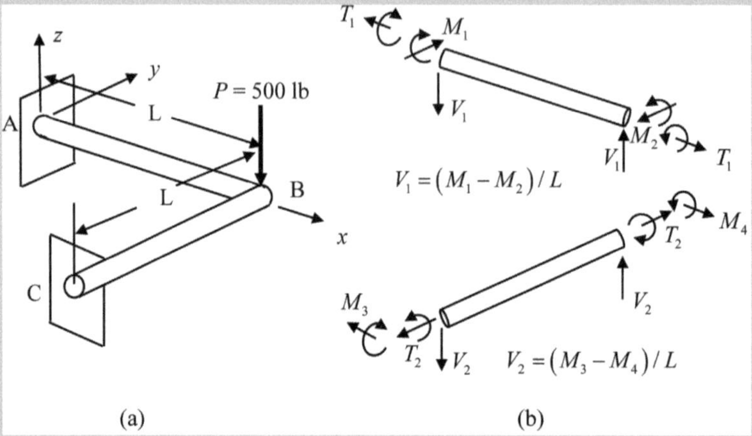

Fig. 5.24 (**a**) A frame under bending and torsion. (**b**) The forces and moments acting on two elements AB and BC

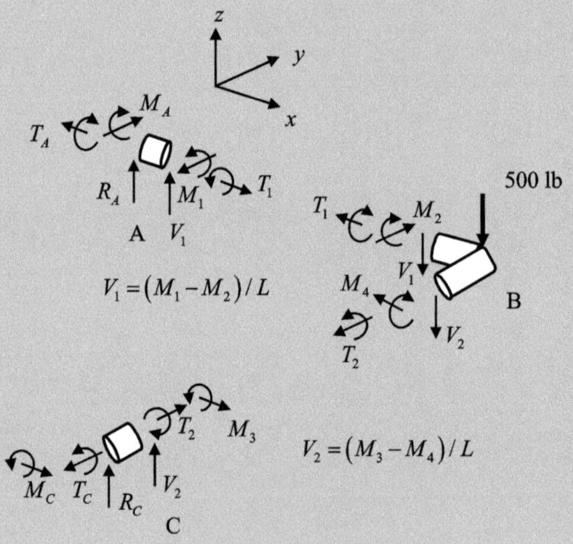

Fig. 5.25 The forces and moments acting at the nodes

(continued)

Example 5.6 (continued)

$$
\begin{aligned}
&{}_A\sum F_z=0 \quad R_A+(M_1-M_2)/L=0\\
&{}_A\sum M_x=0 \quad T_1-T_A=0\\
&{}_A\sum M_y=0 \quad M_A-M_1=0\\
&{}_B\sum F_z=0 \quad -(M_1-M_2)/L-(M_3-M_4)/L-500=0\\
&{}_B\sum M_x=0 \quad -T_1-M_4=0\\
&{}_B\sum M_y=0 \quad M_2-T_2=0\\
&{}_C\sum F_z=0 \quad R_C+(M_3-M_4)/L=0\\
&{}_C\sum M_x=0 \quad M_3+M_C=0\\
&{}_C\sum M_y=0 \quad T_2-T_C=0
\end{aligned}
$$

which are nine equations for twelve unknown forces and moments. In MATLAB, we find.

```
syms Ra Ta Ma Rc Tc Mc M1 M2 M3 M4 T1 T2 EI GJ
L = 4;
Eq(1) = Ra + (M1-M2)/L;
Eq(2) = T1 - Ta;
Eq(3) = Ma - M1;
Eq(4) = -(M1-M2)/L - (M3-M4)/L -500;
Eq(5) = -T1 -M4;
Eq(6) = M2 - T2;
Eq(7) = Rc +(M3-M4)/L;
Eq(8) = M3 + Mc;
Eq(9) = T2 - Tc;
T = [Ra Ta Ma Rc Tc Mc M1 M2 M3 M4 T1 T2];
[E, P] = equationsToMatrix(Eq, T);
S = null(E).';
GJ = 2*EI/3;
gb = (L/EI)*[ 1/3 1/6; 1/6 1/3];
gt = L/GJ;
G = blkdiag(0,0,0,0,0,0,gb,gb,gt, gt);
Esys = [E; S*G];
Psys = [P; 0; 0; 0];
F = Esys\Psys;
Fn = double(F)
```

(continued)

Example 5.6 (continued)

```
Fn =

250     % Ra
-200    % Ta
-800    % Ma
250     % Rc
200     % Tc
800     % Mc
-800    % M1
200     % M2
-800    % M3
200     % M4
-200    % T1
200     % T2
```

The reactions are shown in Fig. 5.26. It is obvious that the forces are in equilibrium. From moment equilibrium about A, we have

$$\sum M_{Ax} = 0 \quad 200 + 800 - (250)(4) = 0$$
$$\sum M_{Ay} = 0 \quad -800 - 200 + (500)(4) - (250)(4) = 0$$

so that equilibrium of moments is also satisfied.

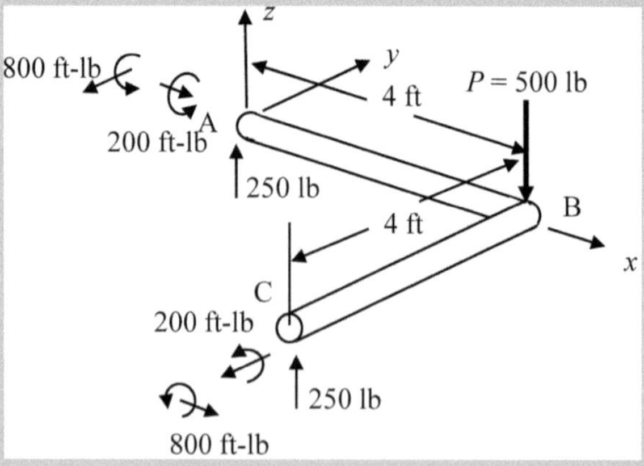

Fig. 5.26 The reactions acting on the frame

5.5 Force-Based Methods for Beams in Strength of Materials

We analyzed beams in this chapter at an advanced strength of materials level because of the many details that are needed in that discussion. However, force-based solutions for beams could be introduced in a strength of materials course using only concepts familiar to the strength of materials students. The differential equations for beams, for example, can be used to justify the homogeneous solutions for bending moment and shear force that form a key part of the force-based method, that is,

$$\begin{aligned} &\frac{d^2M}{dx^2}=0 \\ &\frac{dV}{dx}=0 \quad \rightarrow \quad \begin{aligned} M(x)&=M_1x+(M_2-M_1)x/L \\ V(x)&=(M_1-M_2)/L \end{aligned} \\ &V=-\frac{dM}{dx} \end{aligned} \tag{5.58}$$

Similarly, the equivalent forces/moments acting at the nodes can be justified by solving fixed-fixed problems with the differential equations, as done earlier, so that the bending moment and shear force are then given in each element in the forms shown in Eq. (5.58). Both the flexibility matrix and the deformation-displacement relationship can be obtained from the moment-area theorem since (see Eqs. (5.48) and (5.49))

$$\begin{aligned} t_{B/A}&=\frac{1}{EI}\int_0^L (L-x)M(x)dx \\ t_{A/B}&=\frac{1}{EI}\int_0^L xM(x)dx \end{aligned} \tag{5.59}$$

where $t_{B/A}$ and $t_{A/B}$ are vertical distances between tangents defined previously. From the geometry of an element one can obtain, as shown previously, the relationship between these distances and the nodal displacements and rotations as (see Eqs. (5.50) and (5.51))

$$\begin{aligned} t_{B/A}&=w_2-w_1-\theta_1L \\ t_{A/B}&=w_1-w_2+\theta_2L \end{aligned} \tag{5.60}$$

By integrating Eq. (5.59) and using Eq. (5.60), one obtains the deformation-displacement relations in the form $[G]\{F\}=[E]^T\{-U\}$, or, more explicitly

$$\frac{L}{EI}\begin{bmatrix} 1/3 & 1/6 \\ 1/6 & 1/3 \end{bmatrix}\begin{Bmatrix} M_1 \\ M_2 \end{Bmatrix} = \begin{bmatrix} 1/L & 1 & -1/L & 0 \\ -1/L & 0 & 1/L & -1 \end{bmatrix}\begin{Bmatrix} -w_1 \\ -\theta_1 \\ -w_2 \\ -\theta_2 \end{Bmatrix} \tag{5.61}$$

which, if we let

$$\begin{Bmatrix} \Delta_1 \\ \Delta_2 \end{Bmatrix} = \begin{Bmatrix} t_{B/A}/L \\ t_{A/B}/L \end{Bmatrix}, \quad [G] = \frac{L}{EI}\begin{bmatrix} 1/3 & 1/6 \\ 1/6 & 1/3 \end{bmatrix} \tag{5.62}$$

gives the deformation-displacement relation in the familiar form $\{\Delta\} = [E]^T\{-U\}$, with $[E]$ being the equilibrium matrix for an element given by

$$[E] = \begin{bmatrix} 1/L & -1/L \\ 1 & 0 \\ -1/L & 1/L \\ 0 & -1 \end{bmatrix} \tag{5.63}$$

From these results, we can then implement the complete force-based method since we have found the necessary flexibility matrix associated with the unknowns (M_1, M_2) and the physical meaning of the corresponding deformations and their relationship to the displacements.

5.6 The Connection to Finite Elements

One difference in solving beam problems in the manner we have just illustrated and in solving them by a force-based finite element method is that in finite elements the equilibrium of an element is written in terms of the nodal forces that are the opposite of those used in a node-by-node solution process. Also, the assembly of the global equilibrium matrix is done on an element-by-element basis rather than a node-by-node basis. We discussed these differences for extensional problems in the last chapter. Here we will outline them briefly for beams. Consider, for example, a single element, which we will take as element (1), as shown in Fig. 5.27. The equilibrium equations for this element are, as shown in Fig. (5.27), written as

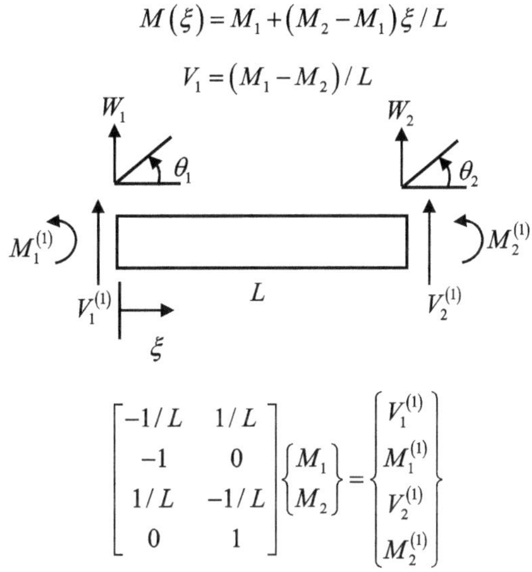

Fig. 5.27 A single element, labeled here as element (1), and the equilibrium equations for the element as written in terms of applied forces/moments acting at the end nodes

$$\begin{bmatrix} -1/L & 1/L \\ -1 & 0 \\ 1/L & -1/L \\ 0 & 1 \end{bmatrix} \begin{Bmatrix} M_1 \\ M_2 \end{Bmatrix} = \begin{Bmatrix} V_1^{(1)} \\ M_1^{(1)} \\ V_2^{(1)} \\ M_2^{(1)} \end{Bmatrix} \tag{5.64}$$

in terms of the nodal forces and moments at the end nodes $\left(V_1^{(1)}, M_1^{(1)}, V_2^{(1)}, M_2^{(1)}\right)$, which are forces/moments acting along the positive axes, which here are the z-axis for the shear forces and the negative y-axis for the bending moments. Contrast this with the equilibrium expression for the single element used in Eq. (5.42) for the problem of Fig. 5.9a, which was

$$\begin{bmatrix} 1/L & -1/L \\ 1 & 0 \\ -1/L & 1/L \\ 0 & -1 \end{bmatrix} \begin{Bmatrix} M_1 \\ M_2 \end{Bmatrix} = \begin{Bmatrix} -V_1^{(1)} \\ -M_1^{(1)} \\ -V_2^{(1)} \\ -M_2^{(1)} \end{Bmatrix} = \begin{Bmatrix} p_0L/2 - V_{R1} \\ p_0L^2/12 - M_{R1} \\ p_0L/2 - V_{R2} \\ -p_0L^2/12 \end{Bmatrix} \tag{5.65}$$

where both the equilibrium matrix and the vector of applied forces are the negatives of those in Eq. (5.64). When we assemble elements in a force-based finite element approach, we simply add elements in the form of Eq. (5.64) in appropriate places in a set of global equilibrium equations. For two identical elements, labeled (1) and (2), for example, we would have (see Fig. 5.28)

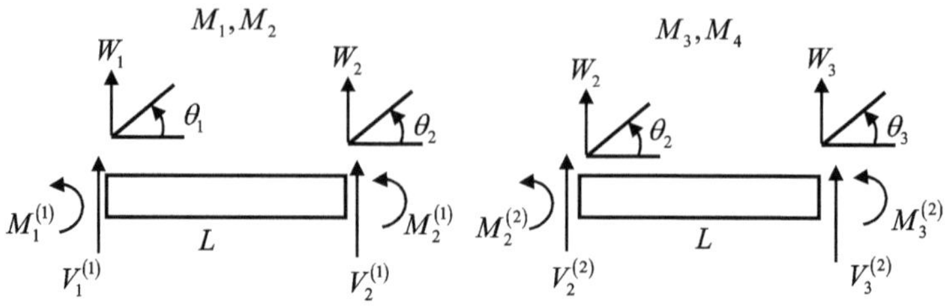

Fig. 5.28 Two finite elements, labeled (1) and (2)

$$\begin{bmatrix} -1/L & 1/L & 0 & 0 \\ -1 & 0 & 0 & 0 \\ 1/L & -1/L & -1/L & 1/L \\ 0 & -1 & -1 & 0 \\ 0 & 0 & 1/L & -1/L \\ 0 & 0 & 0 & 1 \end{bmatrix} \begin{Bmatrix} M_1 \\ M_2 \\ M_3 \\ M_4 \end{Bmatrix} = \begin{Bmatrix} V_1^{(1)} \\ M_1^{(1)} \\ V_2^{(1)} + V_2^{(2)} \\ M_2^{(1)} + M_2^{(2)} \\ V_3^{(2)} \\ M_3^{(2)} \end{Bmatrix} \tag{5.66}$$

The applied forces add appropriately, and the equilibrium matrix is assembled in a very simple fashion since we have taken the elements and their nodes in sequence. In the next chapter, we will discuss how to assemble the global equilibrium matrix when the nodes are nonsequential, as is often the case in more complex problems. Because the force/moment vector on the right side of Eq. (5.66) contains components acting in the positive coordinate directions, if displacements are also calculated, they will also act in those same positive directions and there will be no minus signs appearing in those results, in contrast to the case when we used a node-by-node solution process. This is also true if we use a pseudo stiffness matrix to calculate the displacements. For a single element, the pseudo stiffness matrix is $[K_p] = [E][G]^{-1}[E]^T$, which for a beam element calculated in MATLAB gives.

```
syms L EI
E = [ 1/L -1/L; 1 0; -1/L 1/L; 0 -1];
G = (L/EI)*[1/3 1/6; 1/6 1/3];
Kp = E*inv(G)*E.';
Kp*L^3/EI  % display normalized pseudo stiffness for easier reading

ans =

[ 12,  6*L,  -12,  6*L]
[6*L, 4*L^2, -6*L, 2*L^2]
[-12, -6*L, 12, -6*L]
[6*L, 2*L^2, -6*L, 4*L^2]
```

which is

$$\left[K_p\right] = \frac{EI}{L^3}\begin{bmatrix} 12 & 6L & -12 & 6L \\ 6L & 4L^2 & -6L & 2L^2 \\ -12 & -6L & 12 & -6L \\ 6L & 2L^2 & -6L & 4L^2 \end{bmatrix} \tag{5.67}$$

and for a single element labeled (1) the equilibrium equation in terms of the displacements would be

$$\frac{EI}{L^3}\begin{bmatrix} 12 & 6L & -12 & 6L \\ 6L & 4L^2 & -6L & 2L^2 \\ -12 & -6L & 12 & -6L \\ 6L & 2L^2 & -6L & 4L^2 \end{bmatrix}\begin{Bmatrix} W_1 \\ \theta_1 \\ W_2 \\ \theta_2 \end{Bmatrix} = \begin{Bmatrix} V_1^{(1)} \\ M_1^{(1)} \\ V_2^{(1)} \\ M_2^{(1)} \end{Bmatrix} \tag{5.68}$$

Two identical elements, (1) and (2), when assembled, would yield

$$\frac{EI}{L^3}\begin{bmatrix} 12 & 6L & -12 & 6L & 0 & 0 \\ 6L & 4L^2 & -6L & 2L^2 & 0 & 0 \\ -12 & -6L & 24 & 0 & -12 & 6L \\ 6L & 2L^2 & 0 & 8L^2 & -6L & 2L^2 \\ 0 & 0 & -12 & -6L & 12 & -6L \\ 0 & 0 & 6L & 2L^2 & -6L & 4L^2 \end{bmatrix}\begin{Bmatrix} W_1 \\ \theta_1 \\ W_2 \\ \theta_2 \\ W_3 \\ \theta_3 \end{Bmatrix}$$
$$= \begin{Bmatrix} V_1^{(1)} \\ M_1^{(1)} \\ V_2^{(1)} + V_2^{(2)} \\ M_2^{(1)} + M_2^{(2)} \\ V_3^{(2)} \\ M_3^{(2)} \end{Bmatrix} \tag{5.69}$$

In the stiffness matrix, the overlapping of the element matrices in a global stiffness matrix makes hand assembly rather tedious even when the nodes are numbered sequentially so assembly is best done with a computer program.

In frames, the element equilibrium matrices and the stiffness matrices are even more complex because they can involve extensional, bending, and torsional contributions and they depend on the orientation of the element. However, they are assembled in a similar fashion, a process we will not describe here.

Reference

1. Schmerr, L. W. (2021). *Advanced mechanics of solids*. Cambridge University Press.

Chapter 6
Force-Based and Displacement-Based Finite Elements

Objectives

- To explain the difference between a node-by-node analysis and an element-by-element analysis of statically indeterminate problems.
- To consider force-based and displacement-based finite element methods with one-dimensional examples.
- To describe MATLAB finite element scripts using force-based or displacement-based finite element methods.

6.1 Introduction

Solving equilibrium problems to determine the forces acting in structures has a very long history [1]. Because of the growing capabilities of computers, in the late 1950s and early 1960s, matrix-based methods of analysis became, for the first time, a practical approach. Many of these early applications of matrix methods found a home at Boeing and in the aerospace industry. Civil engineers also had a growing need for more powerful structural analysis tools. The general approach that emerged was to model a complex structure as a collection of elements and then to assemble the equilibrium and deformation behavior of those elements into a large matrix (called a stiffness matrix) that represented the entire structure. Work-energy principles became a key tool for performing this entire process. The fundamental unknowns involved were the displacements of the elements, which (when found) can be used to determine the internal forces everywhere in the structures. Thus, the *finite element method*, as it was first called in 1960, is a displacement-based method that has since grown to be a dominant approach in many fields beyond the structural analysis of statically indeterminate systems, including fluid dynamics, heat transfer, electromagnetics, acoustics, and many more. Earlier structural analysis methods

L. W. Schmerr Jr., *Algebraic Equations of Linear Elasticity*,
https://doi.org/10.1007/978-3-031-66174-7_6

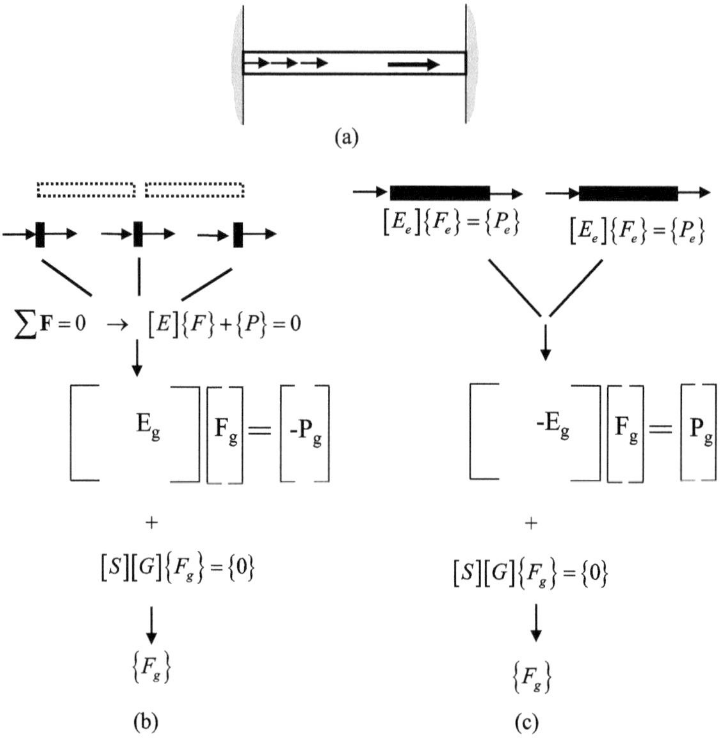

Fig. 6.1 (**a**) An example problem. (**b**) Force-based solution on a node-by-node basis. (**c**) Force-based solution on an element-by-element basis, as used in a finite element method

based on solving for the forces, such as the first-generation force-based method discussed briefly in Chap. 1, have become less frequently taught and used. Second-generation force-based methods, which is the primary topic of this book, can provide a new path to solving statically indeterminate problems, using many of the same steps as found in traditional displacement-based finite element methods. Such force-based finite elements solve directly for the forces rather than displacements using matrix methods and efficient calculations with computers.

In the previous chapters, we have implemented the force-based method on a node-by-node basis that is highly suited to teaching the method in a solid mechanics curriculum. Figure 6.1 illustrates this method and contrasts it to that of a force-based element-by-element finite element approach for a statically indeterminate bar under axial forces. The node-by-node method, as described in detail in Chap. 4 and shown schematically in Fig. 6.1b, begins by representing the bar as a collection of elements. Any applied forces, either concentrated or distributed, are replaced by forces acting at the nodes between elements. Those nodal forces, together with the internal forces from the adjacent elements and reaction forces, if any, are used to define the equilibrium equations at the nodes. The equilibrium equations at each node, which can be written down by hand, are placed symbolically into MATLAB, and

MATLAB can then be used to transform those equations into a global set of matrix-vector equilibrium equations for the unknown forces defined by a global equilibrium matrix $[E_g]$ and a global force vector $\{-P_g\}$. As discussed previously, the minus sign is present because when the applied forces are placed on the right side of the equilibrium equations, $\sum\mathbf{F} = 0$, the vector of known forces will contain the negative of the forces $\{P_g\}$ acting in positive coordinate directions, as shown in Fig. 6.1b. Homogeneous solutions of those global equations are used to define a compatibility matrix and the flexibilities associated with the forces are assembled into a global flexibility matrix. The equilibrium equations and the compatibility equations (which involve the compatibility and flexibility matrices) can be combined into a set of system equations that can then be solved for the forces. Once the forces are known then the displacements, if needed, can be calculated.

In contrast, a force-based finite element method defines an equilibrium matrix $[E_e]$ for each element which is related to a vector of nodal forces $\{P_e\}$ at the ends of the element (see Fig. 6.1c). These element equilibrium matrix and nodal force vectors are assembled into a global equilibrium matrix $[-E_g]$ and a global force vector $\{P_g\}$, where $[E_g]$ is the same as the global matrix obtained on a node-by-node basis. The minus sign is present in this global equilibrium matrix because the force vectors $\{P_g\}$ are always taken to be the forces in the positive coordinate directions acting on the elements (in contrast to the global vector assembled from the node-by-node basis) so the corresponding global equilibrium matrix will be the negative of the one assembled on a node-by-node basis. Note, however, the global equations assembled on either a node-by-node basis or an element-by-element basis yield a completely equivalent set of equations. The remaining steps carry out the solution for the forces (and the displacements). The assembly process in the finite element method can be easily automated on a computer, enabling very large and complex problems to be considered.

The traditional displacement-based finite element method (see Fig. 6.2b) follows an assembly process where element stiffness matrices $[K_e]$ that define the equilibrium of elements in terms of their nodal displacements and nodal forces $\{P_e\}$ are used to assemble a large global stiffness matrix $[K_g]$ for all the elements and a global force vector $\{P_g\}$ for the entire structure. The element stiffness matrix is developed from strain shape functions. After the displacement boundary conditions at the supports are applied, we can solve for the displacements and use them to determine the forces in the elements.

In a displacement-based finite element method based on a pseudo stiffness matrix, the pseudo stiffness matrix is formed from the equilibrium matrix and the inverse of the flexibility matrix, as indicated in Fig. 6.2a. After the displacement boundary conditions associated with reaction forces at the supports are satisfied the displacements can be found and used to determine the forces. In Fig. 6.2 the same force vectors $\{P_e\}$ and $\{P_g\}$ are shown as being used for both methods. If the negative of the applied forces is used in the pseudo stiffness approach, then the negative of the nodal displacements will be obtained in that solution. Note that either an element equilibrium matrix or its negative will produce the same stiffness matrix since both the element equilibrium matrix and its transpose are contained in the definition of the

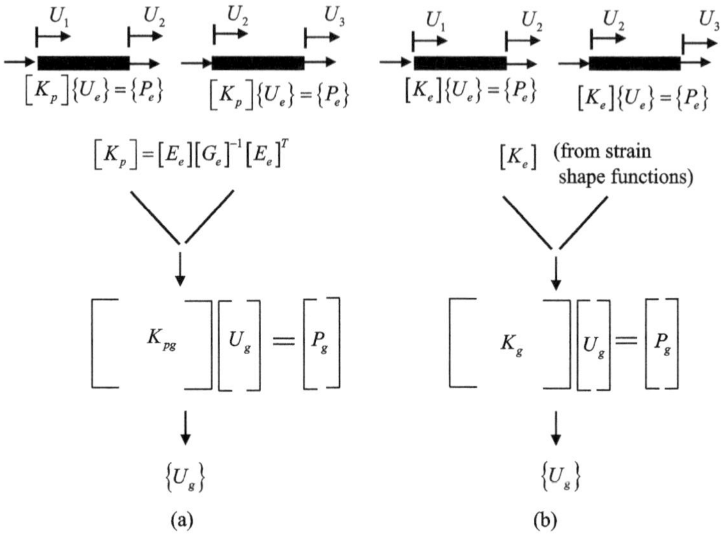

Fig. 6.2 (**a**) Pseudo stiffness-based solution for the displacements in a finite element method. (**b**) The "traditional" finite element method where the element stiffness is defined from displacement (strain) shape functions

stiffness. Thus, the signs on the displacements depend only on whether $\{P_g\}$ are the forces acting in the positive or negative coordinate directions.

When the force-based finite element method is implemented on more complex problems, there can be other differences from the procedures outlined in this book. For example, if many small elements are being used, then any distributed body forces or surface forces may be taken as approximate constants over the elements. Also, while the stresses or forces in elements must always satisfy homogeneous equilibrium equations in the force-based method, this is not true for the displacements. The choices we made for the force and displacement shape functions for extensional, torsion, and bending problems guaranteed that the solutions obtained with the force-based methods for those problems were exact solutions. Other choices can be acceptable in more complex problems but will only generate approximate solutions.

In this chapter, we will discuss both force-based and displacement-based finite elements. Only one-dimensional problems involving axial extension problems of the type considered in Chap. 4 will be used to keep the analysis as simple as possible. An example will be solved without any programming to describe all the steps in the analysis of both force-based and displacement-based methods. Then MATLAB scripts will be developed so that the programming steps needed to implement the methods in a more general 1-D setting can be described.

6.2 Displacement-Based Finite Element Method

Example 6.1: A Fixed-Fixed Bar with Displacement-Based Finite Elements

We will connect our discussion of a displacement-based approach to the specific example shown in Fig. 6.3a, where a bar is fixed at both ends and subjected to a uniform distributed force over its first half and a concentrated force, as shown. The bar will be broken up into six elements (Fig. 6.3b). We could solve this problem exactly with fewer elements, but in a finite element approach, an approximate solution is typically found with many small elements where the shape functions used can be relatively simple, so we will simulate that here. There are six elements and seven nodes that are numbered consecutively, as shown in Fig. 6.3b, where the element numbers are given in parentheses, for example, (n), $n = 1, 2, \ldots 6$). The end reaction forces (R_1, R_2) and the displacements at the nodes are also shown in Fig. 6.3b.

First, we need to define a stiffness matrix for each element. From Chap. 2, we found that the stiffness matrix was given in terms of the strain shape functions $[B]$ and the elastic constants $[C]$ as

$$[K] = \int_V [B]^T [C][B] dV \tag{6.1}$$

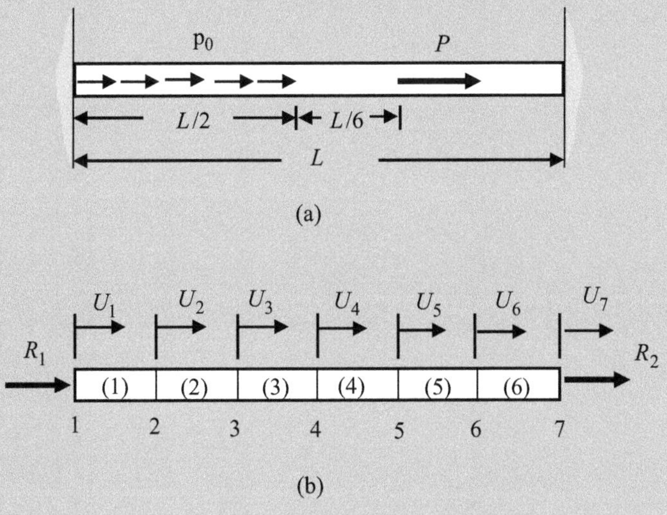

Fig. 6.3 (**a**) A statically indeterminate bar. (**b**) The bar broken into six elements

(continued)

Example 6.1 (continued)

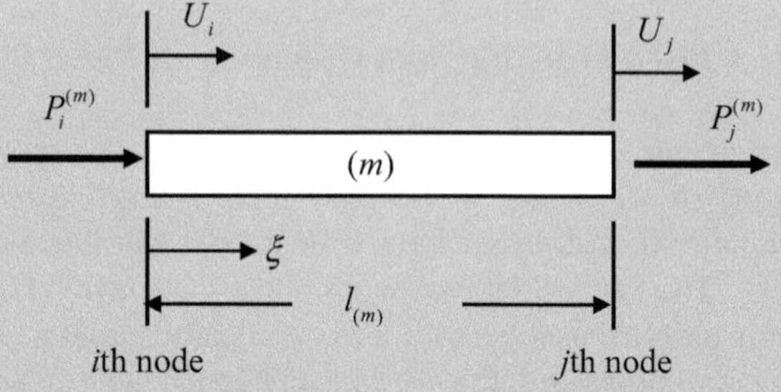

Fig. 6.4 The nodal displacements, end forces, and geometry for the *m*th element

In the extension of a bar, there is only one axial stress σ_{xx} and one axial strain e_{xx}, and they are related through Hooke's law $\sigma_{xx} = Ee_{xx}$, so the elastic constants are just the scalar contant E, which is Young's modulus. For the axial displacement, we will assume a linear relationship in terms of the end nodal displacements (U_i, U_j) for the *m*th element (see Fig. 6.4):

$$u_x^{(m)}(\xi) = U_i + \left(U_j - U_i\right)\xi/l_{(m)} \tag{6.2}$$

where ξ is measured from the start of the element, whose length we will take as $l_{(m)}$. This can be written in terms of displacement shape functions as

$$u_x^{(m)}(\xi) = \left[H_i^{(m)}(\xi) \;\; H_j^{(m)}(\xi)\right]\begin{Bmatrix} U_i \\ U_j \end{Bmatrix} \tag{6.3}$$

where

$$H_i^{(m)}(\xi) = \left(1 - \xi/l_{(m)}\right), \quad H_j^{(m)}(\xi) = \xi/l_{(m)} \tag{6.4}$$

The axial strain is just the derivative of the axial displacement, $e_{xx} = du_x/dx$, so we have

$$e_{xx}^{(m)} = \left[B_i^{(m)} \;\; B_j^{(m)}\right]\begin{Bmatrix} U_i \\ U_j \end{Bmatrix} \tag{6.5}$$

where

$$B_i^{(m)} = -1/l_{(m)}, \quad B_j^{(m)} = 1/l_{(m)} \tag{6.6}$$

which are just constants. If we use all these results to deterrmine the stiffness matrix for an element, we find.

(continued)

Example 6.1 (continued)

$$\left[K^{(m)}\right] = \int_{0}^{l_{(m)}} \begin{bmatrix} -1/l_{(m)} \\ 1/l_{(m)} \end{bmatrix} E^{(m)} \left[-1/l_{(m)} \quad 1/l_{(m)} \right] A^{(m)} d\xi$$
$$= \frac{E^{(m)}A^{(m)}}{l_{(m)}} \begin{bmatrix} 1 & -1 \\ -1 & 1 \end{bmatrix} = \begin{bmatrix} k^{(m)} & -k^{(m)} \\ -k^{(m)} & k^{(m)} \end{bmatrix} \tag{6.7a}$$

where $(E^{(m)}, A^{(m)})$ are Young's modulus and the cross-sectional area for the mth element. The constant $k^{(m)} = E^{(m)}A^{(m)}/l_{(m)}$ is called the stiffness of the element. It acts like a spring constant $k^{(m)}$ for a linear spring. In the problem shown in Fig. 6.3b, *E, A,* and *l* are all constants, so the common stiffness is $k = EA/l$, and the stiffness matrix is the same for all elements:

$$\left[K^{(m)}\right] = \begin{bmatrix} k & -k \\ -k & k \end{bmatrix} \tag{6.7b}$$

Before we can form a set of global equilibrium equations for all the elements, we need to also represent any distributed forces as forces acting at the nodes. In the problem of Fig. 6.3a, there is a distributed force/ unit length $p_x(\xi) = p_0$ acting over the first three elements. As outlined in Chap. 4, we could consider this as a body force that only depends on ξ. In Chap. 2, we found the discrete forces representing a body force *f* as (see Eq. (2.7):

$$\{Q\} = \int_{V} [H]^T \{f\} dV \tag{6.8}$$

so for our one-dimensional axial extension problem, we find the eqiuivalent nodal forces at nodes *i* and *j* for the *m*th element as

$$\{Q\} = \begin{Bmatrix} Q_i^{(m)} \\ Q_j^{(m)} \end{Bmatrix} = \int_{0}^{l_{(m)}} \begin{bmatrix} 1 - \xi/l_{(m)} \\ \xi/l_{(m)} \end{bmatrix} p_x(\xi) A d\xi \tag{6.9}$$

For a uniform force/unit length p_0, we find (see Chap. 4))

$$\begin{Bmatrix} Q_i^{(m)} \\ Q_j^{(m)} \end{Bmatrix} = \begin{Bmatrix} p_0 l_{(m)}/2 \\ p_0 l_{(m)}/2 \end{Bmatrix} \tag{6.10}$$

(continued)

Example 6.1 (continued)

The force P shown in Fig. 6.3a already acts at a node (node five), so it does not have to be replaced.

For each element, the equilibrium equations are (see Eq. (2.5))

$$\left[K^{(m)}\right]\begin{Bmatrix} U_i \\ U_j \end{Bmatrix} = \begin{Bmatrix} P_i^{(m)} \\ P_j^{(m)} \end{Bmatrix} \tag{6.11}$$

where $\left(P_i^{(m)}, P_j^{(m)}\right)$ are the forces acting at the ends of the element (see Fig. 6.4).

We are now in the position to form a global stiffness matrix and write the global equilibrium equations in terms of the displacements for the problem of Fig. 6.3a. Since there are 7 nodes, the global stiffness matrix will be a 7 × 7 matrix. If we place the first element stiffness matrix into the global equilibrium equations, we have

$$\begin{bmatrix} k & -k & 0 & 0 & 0 & 0 & 0 \\ -k & k & 0 & 0 & 0 & 0 & 0 \\ 0 & 0 & 0 & 0 & 0 & 0 & 0 \\ 0 & 0 & 0 & 0 & 0 & 0 & 0 \\ 0 & 0 & 0 & 0 & 0 & 0 & 0 \\ 0 & 0 & 0 & 0 & 0 & 0 & 0 \\ 0 & 0 & 0 & 0 & 0 & 0 & 0 \end{bmatrix} \begin{Bmatrix} U_1 \\ U_2 \\ U_3 \\ U_4 \\ U_5 \\ U_6 \\ U_7 \end{Bmatrix} = \begin{Bmatrix} P_1^{(1)} \\ P_2^{(1)} \\ 0 \\ 0 \\ 0 \\ 0 \\ 0 \end{Bmatrix} \tag{6.12}$$

If we add in the second element contributions, we have

$$\begin{bmatrix} k & -k & 0 & 0 & 0 & 0 & 0 \\ -k & 2k & -k & 0 & 0 & 0 & 0 \\ 0 & -k & k & 0 & 0 & 0 & 0 \\ 0 & 0 & 0 & 0 & 0 & 0 & 0 \\ 0 & 0 & 0 & 0 & 0 & 0 & 0 \\ 0 & 0 & 0 & 0 & 0 & 0 & 0 \\ 0 & 0 & 0 & 0 & 0 & 0 & 0 \end{bmatrix} \begin{Bmatrix} U_1 \\ U_2 \\ U_3 \\ U_4 \\ U_5 \\ U_6 \\ U_7 \end{Bmatrix} = \begin{Bmatrix} P_1^{(1)} \\ P_2^{(1)} + P_2^{(2)} \\ P_3^{(2)} \\ 0 \\ 0 \\ 0 \\ 0 \end{Bmatrix} \tag{6.13}$$

and similarly, adding all the remaining elements, we find

(continued)

Example 6.1 (continued)

$$\begin{bmatrix} k & -k & 0 & 0 & 0 & 0 & 0 \\ -k & 2k & -k & 0 & 0 & 0 & 0 \\ 0 & -k & 2k & -k & 0 & 0 & 0 \\ 0 & 0 & -k & 2k & -k & 0 & 0 \\ 0 & 0 & 0 & -k & 2k & -k & 0 \\ 0 & 0 & 0 & 0 & -k & 2k & -k \\ 0 & 0 & 0 & 0 & 0 & -k & k \end{bmatrix} \begin{Bmatrix} U_1 \\ U_2 \\ U_3 \\ U_4 \\ U_5 \\ U_6 \\ U_7 \end{Bmatrix} = \begin{Bmatrix} P_1^{(1)} \\ P_2^{(1)} + P_2^{(2)} \\ P_3^{(2)} + P_3^{(3)} \\ P_4^{(3)} + P_4^{(4)} \\ P_5^{(4)} + P_5^{(5)} \\ P_6^{(5)} + P_6^{(6)} \\ P_7^{(6)} \end{Bmatrix} \tag{6.14}$$

We now must apply what are called the force boundary conditions, that is, we must specify all the forces acting at the nodes. In our problem, we have

$$\begin{bmatrix} k & -k & 0 & 0 & 0 & 0 & 0 \\ -k & 2k & -k & 0 & 0 & 0 & 0 \\ 0 & -k & 2k & -k & 0 & 0 & 0 \\ 0 & 0 & -k & 2k & -k & 0 & 0 \\ 0 & 0 & 0 & -k & 2k & -k & 0 \\ 0 & 0 & 0 & 0 & -k & 2k & -k \\ 0 & 0 & 0 & 0 & 0 & -k & k \end{bmatrix} \begin{Bmatrix} U_1 \\ U_2 \\ U_3 \\ U_4 \\ U_5 \\ U_6 \\ U_7 \end{Bmatrix} = \begin{Bmatrix} R_1 + p_0 l/2 \\ p_0 l \\ p_0 l \\ p_0 l/2 \\ P \\ 0 \\ R_2 \end{Bmatrix} \tag{6.15}$$

The equivalent forces from the distributed forces acting the first three elements show those equivalent forces adding at nodes two and three, as seen in Eq. (6.15). The reaction forces at the fixed supports must also be added. The global equations are now complete, but they cannot be solved at this stage since we have not ensured that the displacements at the fixed supports are zero (i.e., enforcing what are called the displacement boundary conditions). One way to proceed to note that since $U_1 = U_7 = 0$, we can remove them entirely from the problem by eliminating the first and seventh rows and columns in the global stiffness matrix and the first and seventh components of $\{U_g\}$ and $\{P_g\}$, giving

$$\begin{bmatrix} 2k & -k & 0 & 0 & 0 \\ -k & 2k & -k & 0 & 0 \\ 0 & -k & 2k & -k & 0 \\ 0 & 0 & -k & 2k & -k \\ 0 & 0 & 0 & -k & 2k \end{bmatrix} \begin{Bmatrix} U_2 \\ U_3 \\ U_4 \\ U_5 \\ U_6 \end{Bmatrix} = \begin{Bmatrix} p_0 l \\ p_0 l \\ p_0 l/2 \\ P \\ 0 \end{Bmatrix} \tag{6.16}$$

which we will write in matrix-vector form as $[K]\{U_f\} = \{P\}$, where $[K]$ is the stiffness matrix for the reduced system, $\{U_f\}$ is a vector of all the unknown

(continued)

Example 6.1 (continued)
"free" displacements, and $\{P\}$ is a vector of all the known forces. We can solve this system of linear equations in MATLAB. To do so, we must specify the parameters. Let $k = 4 \times 10^6$ lb/in, $p_0 = 100$ lb/in, $l = 6$ inches, $P = 50$ lb. Since we have already assembled the stiffness matrix by hand, we can form this matrix by placing all the elements of the main diagonal in a vector, D, the elements above the main diagonal in a vector U, and the elements below the diagonal in a vector L and then use the MATLAB diag function to form the full matrix. Later, we will show how to assemble the global stiffness matrix with a set of more general programming steps. Here is the MATLAB solution:

```
k = 4*10^6; p0 = 100; l = 6; P = 600; % give parameters
U = -k*ones(1,4); % form up stiffness matrix and show it
L = -k*ones(1,4);
D = 2*k*ones(1,5);
K = diag(D,0) + diag(U,1) + diag(L, -1)
K =

   8000000   -4000000         0          0          0
  -4000000    8000000  -4000000          0          0
         0   -4000000   8000000   -4000000          0
         0          0  -4000000    8000000   -4000000
         0          0         0   -4000000    8000000

Pv = [p0*l; p0*l; p0*l/2; P; 0]; % force vector
Uf = K\Pv

Uf =

 1.0e-03 *

 0.3125    % U2
 0.4750    % U3
 0.4875    % U4
 0.4250    % U5
 0.2125    % U6
```

If we then put the zero end displacements back into a global set of displacements, Ug, we can solve for the forces (which are constants) in the elements since the forces are $F^{(m)} = k^{(m)}(U_j - U_i)$ (see Fig. 6.4):

(continued)

Example 6.1 (continued)

```
Ug = [0;Uf;0];          % global displacements
Ue = Ug(2:end);         % end displacements of the elements
Ub = Ug(1:end-1);       % beginning displacements of the elements
F = k*(Ue - Ub)         % the forces in the six elements (lb)
F =

 1.0e+03 *

 1.2500  % F1
 0.6500  % F2
 0.0500  % F3
 -0.2500  % F4
 -0.8500  % F5
 -0.8500  % F6
```

We can also determine the reactions by going back to the first and last global equations in Eq. (6.15) and putting in the known global displacements present in those equations:

```
R1 = -k*Ug(2) - p0*l/2 % Ug(1) = 0 so need not include it
R1 =

 -1.5500e+03

R2 = -k*Ug(6) % Ug(7) = 0 so need not include it

R2 =

 -850.0000
```

so that we have R1 = −1550 lb. and R2 = −850 lb. which you can verify are in equilibrium with the applied forces.

We now would like to plot the forces and compare them with the exact solution, which we can express using singularity functions (see Chap. 4). Using the reactions and the applied forces, we can write an anonymous function to evaluate and plot the internal force in the bar. We can then compare those exact results with the forces obtained from the finite element solution. Since the finite element forces are constant through the element while the exact forces will vary in an element (at least when within elements where the distributed force exists), we will plot the finite element forces at the centroids of each element:

(continued)

Example 6.1 (continued)

```
x = linspace(0, 36, 1000);
Force = @(x) 1550*step_sf(x,0) -100*lin_sf(x,0) + 100*lin_sf(x,
18) ...
 - 600*step_sf(x,24) + 850*step_sf(x,36);
plot(x, Force(x))
hold on
xc = [3 9 15 21 27 33];  % centroids of the elements
plot(xc, F, 'bo')        % plot forces in elements
hold off
```

The results are shown in Fig. 6.5, where the labels have been added separately. We see the values of the reaction forces at the fixed ends and the jump in the internal force at the concentrated force P. Obviously, we will need many more elements to capture that jump behavior with just the nodal values. Since the displacement is related to an integral of the force given by (see Eq. (4.22)):

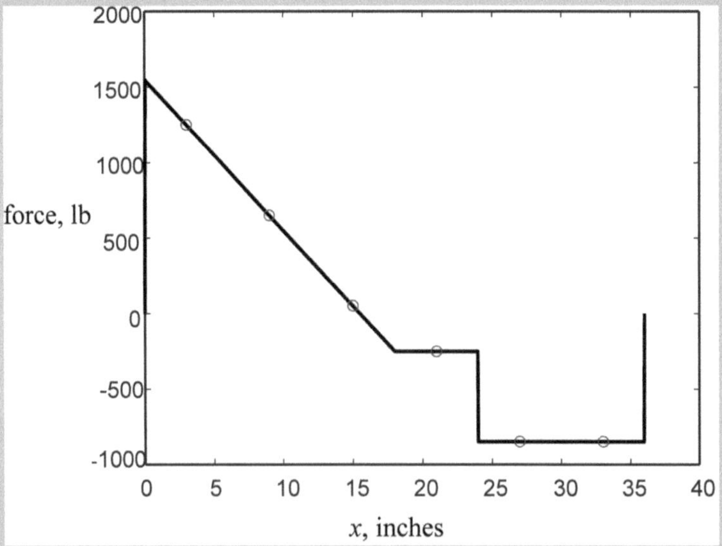

Fig. 6.5 The exact force distribution in the bar (solid line) and the discrete finite element forces (open circles) plotted at the centers of the elements

(continued)

Example 6.1 (continued)

$$u(x) = \frac{1}{EA}\int_0^x F(x)dx = \frac{1}{kl}\int_0^x F(x)dx \tag{6.17}$$

we can define an anonymous displacement function as an integral of the force function and compare it to the global nodal displacements:

```
Displ = @(x) (1/(k*l))*( 1550*lin_sf(x,0) -50*quad_sf(x,0) +
50*quad_sf(x, 18) ...
 - 600*lin_sf(x,24) + 850*lin_sf(x,36));
plot(x, Displ(x))
hold on
xn = [ 0 6 12 18 24 30 36];   % locations of the nodes ( in inches)
plot(xn, Uq, 'bo')
hold off
```

The results are shown in Fig. 6.6, where again the labels have been added separately. There is exact agreement at the nodes, but in the elements where the distributed force exists, the exact displacement is a quadratic function, while our finite element solution assumes linear behavior between the nodes. Of course, more elements can lead to a better solution there.

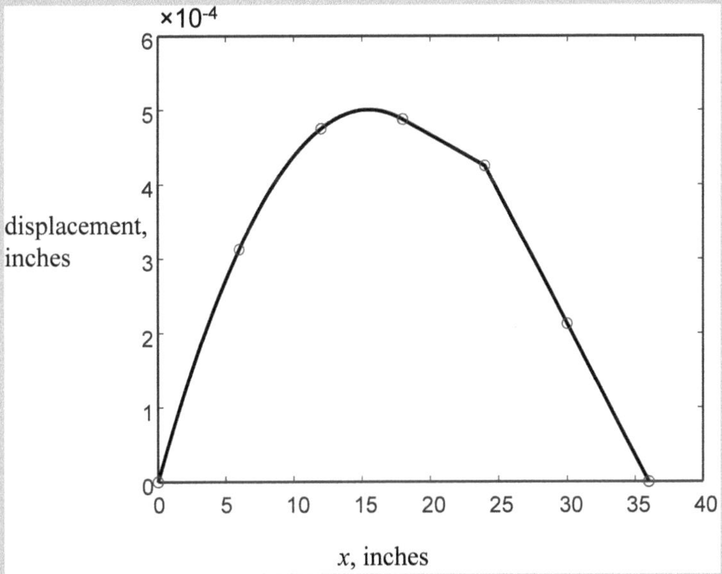

Fig. 6.6 The exact displacement distribution in the bar (solid line) and the nodal displacements (open circles)

(continued)

Example 6.1 (continued)

As we have seen previously, there is another way to satisfy the boundary conditions that does not involve eliminating equations from the global equilibrium equations but instead modifies those global equations. This is often the procedure used in finite elements. Let's go back to Eq. (6.14), which we rewrite here:

$$\begin{bmatrix} k & -k & 0 & 0 & 0 & 0 & 0 \\ -k & 2k & -k & 0 & 0 & 0 & 0 \\ 0 & -k & 2k & -k & 0 & 0 & 0 \\ 0 & 0 & -k & 2k & -k & 0 & 0 \\ 0 & 0 & 0 & -k & 2k & -k & 0 \\ 0 & 0 & 0 & 0 & -k & 2k & -k \\ 0 & 0 & 0 & 0 & 0 & -k & k \end{bmatrix} \begin{Bmatrix} U_1 \\ U_2 \\ U_3 \\ U_4 \\ U_5 \\ U_6 \\ U_7 \end{Bmatrix} = \begin{Bmatrix} P_1^{(1)} \\ P_2^{(1)} + P_2^{(2)} \\ P_3^{(2)} + P_3^{(3)} \\ P_4^{(3)} + P_4^{(4)} \\ P_5^{(4)} + P_5^{(5)} \\ P_6^{(5)} + P_6^{(6)} \\ P_7^{(6)} \end{Bmatrix} \tag{6.18}$$

We first want to satisfy the force conditions, which involve placing all the known forces in the force vector of Eq. (6.18), and then we want to satisfy the displacement boundary conditions, which here are the conditions $U_1 = U_7 = 0$ at the fixed ends. There are three steps:

Step 1. Initialize the right-hand side force vector to be all zeros

$$\{P_g\} = \begin{Bmatrix} 0 \\ 0 \\ 0 \\ 0 \\ 0 \\ 0 \\ 0 \end{Bmatrix} \tag{6.19}$$

Step 2. Place all the known forces at the nodes which do not involve the displacements. In the present case, we find

$$\{P_g\} = \begin{Bmatrix} 0 \\ p_0 l \\ p_0 l \\ p_0 l/2 \\ P \\ 0 \\ 0 \end{Bmatrix} \tag{6.20}$$

(continued)

Example 6.1 (continued)

Step 3. If a displacement is specified as zero at the nth node, then leave the right-hand side force value at that node as zero and multiply the K_{nn} term in the diagonal of the global stiffness matrix by a large value N. For example, choosing $N = 10^6$, we would have

$$\begin{bmatrix} 10^6k & -k & 0 & 0 & 0 & 0 & 0 \\ -k & 2k & -k & 0 & 0 & 0 & 0 \\ 0 & -k & 2k & -k & 0 & 0 & 0 \\ 0 & 0 & -k & 2k & -k & 0 & 0 \\ 0 & 0 & 0 & -k & 2k & -k & 0 \\ 0 & 0 & 0 & 0 & -k & 2k & -k \\ 0 & 0 & 0 & 0 & 0 & -k & 10^6k \end{bmatrix} \begin{Bmatrix} U_1 \\ U_2 \\ U_3 \\ U_4 \\ U_5 \\ U_6 \\ U_7 \end{Bmatrix} = \begin{Bmatrix} 0 \\ p_0 l \\ p_0 l \\ p_0 l/2 \\ P \\ 0 \\ 0 \end{Bmatrix} \tag{6.21}$$

In Eq. (6.21), the first and last equations are simply the zero displacement boundary conditions $U_1 = U_7 = 0$ to a very good approximation. The reaction forces are now missing in those equations since we are using the first and seventh equations to represent the displacement boundary conditions at the nodes where those reactions act. Both the force and displacement boundary conditions are now satisfied in Eq. (6.21), and we can solve these equations for the global displacements. In the more general case where a displacement is specified as a nonzero value, C, at the nth node, we replace the right-hand side value in the force vector at that node by the value $CK_{nn}N$, where N is again a large value and we replace the K_{nn} term in the stiffness matrix by NK_{nn}. In the present example being considered, for example, if we had the displacement boundary condition $U_1 = .003$ in instead of $U_1 = 0$, the modified equations would be

$$\begin{bmatrix} 10^6k & -k & 0 & 0 & 0 & 0 & 0 \\ -k & 2k & -k & 0 & 0 & 0 & 0 \\ 0 & -k & 2k & -k & 0 & 0 & 0 \\ 0 & 0 & -k & 2k & -k & 0 & 0 \\ 0 & 0 & 0 & -k & 2k & -k & 0 \\ 0 & 0 & 0 & 0 & -k & 2k & -k \\ 0 & 0 & 0 & 0 & 0 & -k & 10^6k \end{bmatrix} \begin{Bmatrix} U_1 \\ U_2 \\ U_3 \\ U_4 \\ U_5 \\ U_6 \\ U_7 \end{Bmatrix} = \begin{Bmatrix} .003 \times 10^6k \\ p_0 l \\ p_0 l \\ p_0 l/2 \\ P \\ 0 \\ 0 \end{Bmatrix} \tag{6.22}$$

which would satisfy to a very good approximation $U_1 = .003$, $U_7 = 0$. Once again, we could then solve these modified equations. Consider our previous problem in MATLAB using these procedures (with zero displacements at the ends):

(continued)

Example 6.1 (continued)

```
k = 4*10^6; p0 = 100; l = 6; P = 600; % give parameters
```

When we form the global stiffness matrix, we need to modify some diagonal terms (see Eq. (6.14)).

```
U = -k*ones(1,6);          % form up global stiffness matrix
L = -k*ones(1,6);
D = 2*k*ones(1,7);
Kg = diag(D,0) + diag(U,1) + diag(L, -1);
Kg(1,1) = Kg(1,1)/2;      % modify these terms to obtain correct
stiffness matrix
Kg(7,7) = Kg(7,7)/2;
```

Here is the force vector and stiffness matrix after applying force and displacement boundary conditions:

```
Pg = [ 0; p0*l; p0*l; p0*l/2; P; 0; 0];    % force vector
Kg(1,1) = Kg(1,1)*10^6;
Kg(7,7) = Kg(7,7)*10^6;
```

Now solve for the global displacements:

```
Ug = Kg\Pg
Ug =

 1.0e-03 *

 0.0000   % U1
 0.3125   % U2
 0.4750   % U3
 0.4875   % U4
 0.4250   % U5
 0.2125   % U6
 0.0000   % U7
```

and find the forces, as before:

```
Ue = Ug(2:end);   % end displacements of the elements
Ub = Ug(1:end-1);   % beginning displacements of the elements
F = k*(Ue - Ub)
```

(continued)

Example 6.1 (continued)

```
F = 1.0e+03 *
 1.2500   % F1
 0.6500   % F2
 0.0500   % F3
 -0.2500   % F4
 -0.8500   % F5
 -0.8500   % F6
```

These results look identical to our previous results, but if we display more decimal points in MATLAB, we see that the displacements are not exact, but a very good approximation indeed:

```
format long
Ug
Ug =
  1.0e-03 *
  0.000000312500296
  0.312500295833618
  0.475000279166940
  0.487500262500263
  0.425000245833585
  0.212500229166907
  0.000000212500229
```

We can also find the reactions at the fixed supports as done before.

6.3 Force-Based Finite Element Method

The first step in the force-based finite element method is to find the equilibrium matrix $[E^{(m)}]$ for an element. For the axial extension of a bar, we can use the general form we developed in Chap. 2 (see Eq. (2.16)):

$$\left[E^{(m)}\right] = \int_V [B]^T [N] dV \tag{6.23}$$

where $[B]$ and $[N]$ are shape functions for the strain and stress in the bar. The stress shape functions, however, must satisfy the homogeneous equilibrium equations which for a bar, where there is only one stress, σ_{xx}, must satisfy $d\sigma_{xx}/dx = 0$, so the stress must just be a constant σ_0 and the shape function $[N] = 1$. However, we are using the force $F = \sigma_0 A$ in our equilibrium equations in place of the stress so we have instead $[N] = 1/A$. From Hooke's law, $\sigma_{xx} = Ee_{xx}$, so the strain must also be a constant and, therefore, the displacement in the bar must be a linear function of ξ, as found before. We have already written down the shape function for the strain as (see Eqs. (6.5) and (6.6))

$$[B] = \begin{bmatrix} -1/l_{(m)} & 1/l_{(m)} \end{bmatrix} \tag{6.24}$$

so the element equilibrium matrix is

$$\left[E^{(m)}\right] = \int_0^{l_{(m)}} \left(\frac{1}{A}\right) \begin{bmatrix} -1/l_{(m)} \\ 1/l_{(m)} \end{bmatrix} A d\xi = \begin{bmatrix} -1 \\ 1 \end{bmatrix} \tag{6.25}$$

The internal force $\{F_m\} = F_m = \sigma_0 A$ is a constant force in the mth element of the bar (assuming A is not a function of ξ). The element equilibrium equations, therefore, are $[E^{(m)}]\{F_m\} = \{P^{(m)}\}$, or more explicitly (see Fig. 6.7):

$$\begin{bmatrix} -1 \\ 1 \end{bmatrix} \{F_m\} = \begin{Bmatrix} P_i^{(m)} \\ P_j^{(m)} \end{Bmatrix} \tag{6.26}$$

From Eq. (6.26) and Fig. 6.7c, it follows that these element equilibrium equations are just statements of equilibrium between the internal force F_m and the forces at the end nodes of the element, where the force vectors are acting in the positive x-direction.

Using the problem of Fig. 6.3a as an example, we can assemble the global equilibrium equations element-by element for the six elements and seven nodes, replace the distributed forces with equivalent nodal forces, and arrive at a global

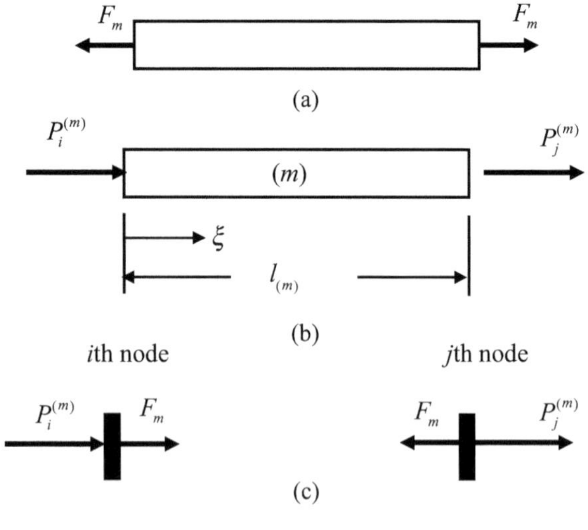

Fig. 6.7 (**a**) The internal force in the mth element of a bar. (**b**) The end forces acting on the bar element geometry. (**c**) The forces at the end nodes of the element

system $[E_g]\{F\} = \{P_g\}$. The steps are very similar to the displacement case, so we just write down the result for the global system:

$$\begin{bmatrix} -1 & 0 & 0 & 0 & 0 & 0 \\ 1 & -1 & 0 & 0 & 0 & 0 \\ 0 & 1 & -1 & 0 & 0 & 0 \\ 0 & 0 & 1 & -1 & 0 & 0 \\ 0 & 0 & 0 & 1 & -1 & 0 \\ 0 & 0 & 0 & 0 & 1 & -1 \\ 0 & 0 & 0 & 0 & 0 & 1 \end{bmatrix} \begin{Bmatrix} F_1 \\ F_2 \\ F_3 \\ F_4 \\ F_5 \\ F_6 \end{Bmatrix} = \begin{Bmatrix} P_1^{(1)} \\ P_2^{(1)} + P_2^{(2)} \\ P_3^{(2)} + P_3^{(3)} \\ P_4^{(3)} + P_4^{(4)} \\ P_5^{(4)} + P_5^{(5)} \\ P_6^{(5)} + P_6^{(6)} \\ P_7^{(6)} \end{Bmatrix} = \begin{Bmatrix} R_1 + p_0 l/2 \\ p_0 l \\ p_0 l \\ p_0 l/2 \\ P \\ 0 \\ R_2 \end{Bmatrix} \tag{6.27}$$

This is seven equations in eight unknowns when we include the reaction forces (R_1, R_2). Even if we eliminate the first and last equations that contain the reactions, we have five equations in six unknowns. This shows that our problem is indeed statically indeterminate. To find the other equation(s) needed, recall the internal forces in the elements are related to the internal elongations, Δ_m, of the elements, through $\Delta_m = g_m F_m$ where $g_m = 1/k_m$ is the *flexibility*. The elongation of the *m*th bar element, Δ_m, is the difference between the end displacements, so we have

$$\Delta_m = [\,-1 \quad 1\,]\begin{Bmatrix} U_i \\ U_j \end{Bmatrix} \tag{6.28}$$

We can assemble these relations for all the elements. For the problem being discussed, we have

$$\begin{Bmatrix} \Delta_1 \\ \Delta_2 \\ \Delta_3 \\ \Delta_4 \\ \Delta_5 \\ \Delta_6 \end{Bmatrix} = \begin{bmatrix} -1 & 1 & 0 & 0 & 0 & 0 & 0 \\ 0 & -1 & 1 & 0 & 0 & 0 & 0 \\ 0 & 0 & -1 & 1 & 0 & 0 & 0 \\ 0 & 0 & 0 & -1 & 1 & 0 & 0 \\ 0 & 0 & 0 & 0 & -1 & 1 & 0 \\ 0 & 0 & 0 & 0 & 0 & -1 & 1 \end{bmatrix} \begin{Bmatrix} U_1 \\ U_2 \\ U_3 \\ U_4 \\ U_5 \\ U_6 \\ U_7 \end{Bmatrix} \tag{6.29}$$

which we can write in terms of the global equilibrium matrix as $\{\Delta\} = [E_g]^T\{U_g\}$, where T denotes the matrix transpose. Thus, these elongations are the deformations that appear in the algebraic equations of linear elasticity and the deformations are related to the internal forces in the entire bar through $\{\Delta\} = [G]\{F\}$, where $[G]$ is the 6×6 flexibility matrix

$$[G] = \begin{bmatrix} 1/k_1 & 0 & 0 & 0 & 0 & 0 \\ 0 & 1/k_2 & 0 & 0 & 0 & 0 \\ 0 & 0 & 1/k_3 & 0 & 0 & 0 \\ 0 & 0 & 0 & 1/k_4 & 0 & 0 \\ 0 & 0 & 0 & 0 & 1/k_5 & 0 \\ 0 & 0 & 0 & 0 & 0 & 1/k_6 \end{bmatrix} \tag{6.30}$$

assuming the flexibilities of each element are different.

Now, consider how to solve our problem. One way is to eliminate reactions from the global equations by deleting the first and seventh rows in the global equilibrium matrix and the first and last terms in the force vector, yielding a set of "free" equations $[E_f]\{F\} = \{P_f\}$ given by

$$\begin{bmatrix} 1 & -1 & 0 & 0 & 0 & 0 \\ 0 & 1 & -1 & 0 & 0 & 0 \\ 0 & 0 & 1 & -1 & 0 & 0 \\ 0 & 0 & 0 & 1 & -1 & 0 \\ 0 & 0 & 0 & 0 & 1 & -1 \end{bmatrix} \begin{Bmatrix} F_1 \\ F_2 \\ F_3 \\ F_4 \\ F_5 \\ F_6 \end{Bmatrix} = \begin{Bmatrix} p_0 l \\ p_0 l \\ p_0 l/2 \\ P \\ 0 \end{Bmatrix} \tag{6.31}$$

Similarly, if we delete the first and seventh columns of the matrix in the deformation-displacement relations, we find

$$\begin{Bmatrix} \Delta_1 \\ \Delta_2 \\ \Delta_3 \\ \Delta_4 \\ \Delta_5 \\ \Delta_6 \end{Bmatrix} = \begin{bmatrix} 1 & 0 & 0 & 0 & 0 \\ -1 & 1 & 0 & 0 & 0 \\ 0 & -1 & 1 & 0 & 0 \\ 0 & 0 & -1 & 1 & 0 \\ 0 & 0 & 0 & -1 & 1 \\ 0 & 0 & 0 & 0 & -1 \end{bmatrix} \begin{Bmatrix} U_2 \\ U_3 \\ U_4 \\ U_5 \\ U_6 \end{Bmatrix} \tag{6.32}$$

which we can write as $\{\Delta\} = [E_f]^T\{U_f\}$, where $\{U_f\}$ are the "free" displacements in the problem. In summary, at this stage, we have the free equilibrium equations, the deformation-free displacement relations, and the deformation-force relations:

$$[E_f]\{F\} = \{P_f\}, \quad \{\Delta\} = [E_f]^T\{U_f\}, \quad \{\Delta\} = [G]\{F\} \tag{6.33}$$

These are just a special case of the algebraic equations of linear elasticity obtained in Chap. 2 and can be solved in the same fashion. Here we have five equilibrium equations for six forces, six deformation-displacement relations for six deformations and five displacements, six deformation-force relations for a total of 17 equations and 17 unknowns. To solve this set of equations by a force-based method, therefore, we find the compatibility matrix $[S]$ that defines the compatibility equations $[S]\{\Delta\} = \{0\}$ between the deformations, which here is just a single equation. As seen in Chap. 2, this compatibility matrix can be found from homogeneous solutions of the free equilibrium equations, that is,

$$[E_f][S]^T = [0] \tag{6.34}$$

and the compatibility equations can then be written in terms of the forces as

$$[S][G]\{F\} = \{0\} \tag{6.35}$$

Combining the free equilibrium equations with these compatibility equations, we can define a set of "system" equations:

$$\begin{bmatrix} [E_f] \\ - \\ [S][G] \end{bmatrix} \{F\} = \begin{Bmatrix} \{P_f\} \\ - \\ \{0\} \end{Bmatrix} \tag{6.36}$$

which we will write as $[E_{sys}]\{F\} = \{P_{sys}\}$. In the present problem, these are six equations for the six unknown forces in the elements, which can be solved for those forces. As shown in Chap. 2, once the forces have been obtained, we can find the free displacements $\{U_f\}$ by solving the equations

$$[E_{sys}]^T \begin{Bmatrix} \{U_f\} \\ - \\ \{0\} \end{Bmatrix} = [G]\{F\} \tag{6.37}$$

which we will write as

$$[E_{sys}]^T\{U_{sys}\} = [G]\{F\} \tag{6.38}$$

where $\{U_{sys}\}$ contains the free displacements and a number of zeros equal to the number of compatibility equations (which is one in the present case). Note that we obtain the displacements, not the negative of the displacements because the forces contained in the force vector act in the positive x-direction.

Example 6.2: A Fixed-Fixed Bar with Force-based Finite Elements
Now consider solving the problem of Fig. 6.3a with a force-based finite element method.

First, we give the same parameters as before and include the flexibility.

(continued)

Example 6.2 (continued)

```
k = 4*10^6; p0 = 100; l = 6; P = 600;      % give parameters
g = 1/k;     % flexibility
```

Next, we will assemble the free equilibrium matrix and force vector by hand:

```
Ef = [1 -1 0 0 0 0; 0 1 -1 0 0 0; 0 0 1 -1 0 0 ...
   ; 0 0 0 1 -1 0; 0 0 0 0 1 -1];
Pf = [ p0*l; p0*l; p0*l/2; P; 0];
```

Then we will solve for the compatibility matrix and form the flexibility matrix:

```
S = null(Ef).';
flex = g*ones(1,6);
G = diag(flex);
```

We will define the system equations and solve them for the forces:

```
Esys = [Ef; S*G];

Psys = [ Pf;0];
F = Esys\Psys

F =

 1.0e+03 *

 1.2500 % F1
 0.6500 % F2
 0.0500 % F3
 -0.2500% F4
 -0.8500% F5
 -0.8500% F6
```

and then obtain the free displacements (and an additional zero):

```
Uf = Esys.'\G*F
Uf =
```

(continued)

Example 6.2 (continued)

```
1.0e-03 *

0.3125 % U2
0.4750 % U3
0.4875 % U4
0.4250 % U5
0.2125 % U6
-0.0000% extra zero
```

You can see that these are the same values obtained with the displacement-based method.

There is another way to solve the problem with a force-based approach. In this case, we expand the global equilibrium matrix by bringing the unknown reactions into those equations. Equation (6.27) then becomes

$$\begin{bmatrix} -1 & 0 & 0 & 0 & 0 & 0 & -1 & 0 \\ 1 & -1 & 0 & 0 & 0 & 0 & 0 & 0 \\ 0 & 1 & -1 & 0 & 0 & 0 & 0 & 0 \\ 0 & 0 & 1 & -1 & 0 & 0 & 0 & 0 \\ 0 & 0 & 0 & 1 & -1 & 0 & 0 & 0 \\ 0 & 0 & 0 & 0 & 1 & -1 & 0 & 0 \\ 0 & 0 & 0 & 0 & 0 & 1 & 0 & -1 \end{bmatrix} \begin{Bmatrix} F_1 \\ F_2 \\ F_3 \\ F_4 \\ F_5 \\ F_6 \\ R_1 \\ R_2 \end{Bmatrix} = \begin{Bmatrix} p_0 l/2 \\ p_0 l \\ p_0 l \\ p_0 l/2 \\ P \\ 0 \\ 0 \end{Bmatrix} \quad (6.39)$$

which we will write as the "extended" global equations $[E_{ge}]\{F_e\} = \{P_e\}$. The deformation and displacement relations in terms of this extended matrix is

$$\begin{Bmatrix} \Delta_1 \\ \Delta_2 \\ \Delta_3 \\ \Delta_4 \\ \Delta_5 \\ \Delta_6 \\ \Delta_{R1} \\ \Delta_{R2} \end{Bmatrix} = \begin{bmatrix} -1 & 1 & 0 & 0 & 0 & 0 & 0 \\ 0 & -1 & 1 & 0 & 0 & 0 & 0 \\ 0 & 0 & -1 & 1 & 0 & 0 & 0 \\ 0 & 0 & 0 & -1 & 1 & 0 & 0 \\ 0 & 0 & 0 & 0 & -1 & 1 & 0 \\ 0 & 0 & 0 & 0 & 0 & -1 & 1 \\ -1 & 0 & 0 & 0 & 0 & 0 & 0 \\ 0 & 0 & 0 & 0 & 0 & 0 & -1 \end{bmatrix} \begin{Bmatrix} U_1 \\ U_2 \\ U_3 \\ U_4 \\ U_5 \\ U_6 \\ U_7 \end{Bmatrix} \quad (6.40)$$

where deformations associated with the reaction forces have been added. This we will write as $\{\Delta_e\} = [E_{ge}]^T\{U_g\}$. We must set these additional deformations to zero, that is, $\Delta_{R1} = \Delta_{R2} = 0$. Equation (6.40) shows that then we will

(continued)

Example 6.2 (continued)
also have $\Delta_{R1} = -U_1 = 0$ and $\Delta_{R2} = -U_7 = 0$, so we will satisfy the displacement boundary conditions at the fixed supports. To ensure the added deformations are zero, we must associate zero flexibilities with those deformations, and the flexibility matrix in the deformation-force relations will be

$$\begin{Bmatrix} \Delta_1 \\ \Delta_2 \\ \Delta_3 \\ \Delta_4 \\ \Delta_5 \\ \Delta_6 \\ \Delta_{R1} \\ \Delta_{R2} \end{Bmatrix} = \begin{bmatrix} 1/k & 0 & 0 & 0 & 0 & 0 & 0 & 0 \\ 0 & 1/k & 0 & 0 & 0 & 0 & 0 & 0 \\ 0 & 0 & 1/k & 0 & 0 & 0 & 0 & 0 \\ 0 & 0 & 0 & 1/k & 0 & 0 & 0 & 0 \\ 0 & 0 & 0 & 0 & 1/k & 0 & 0 & 0 \\ 0 & 0 & 0 & 0 & 0 & 1/k & 0 & 0 \\ 0 & 0 & 0 & 0 & 0 & 0 & 0 & 0 \\ 0 & 0 & 0 & 0 & 0 & 0 & 0 & 0 \end{bmatrix} \begin{Bmatrix} F_1 \\ F_2 \\ F_3 \\ F_4 \\ F_5 \\ F_6 \\ R_1 \\ R_2 \end{Bmatrix} \tag{6.41}$$

which will be written as $\{\Delta_e\} = [G_e]\{F_e\}$. We can determine the compatibility matrix from the homogeneous solutions $[E_{ge}][S]^T = [0]$, and then the rest of the solution for the forces (including the reactions) becomes

$$\begin{bmatrix} [E_{ge}] \\ - \\ [S][G_e] \end{bmatrix} \{F_e\} = \begin{Bmatrix} \{P_e\} \\ - \\ \{0\} \end{Bmatrix} \tag{6.42}$$

which we write again as $[E_{sys}]\{F_e\} = \{P_{sys}\}$, and the displacements are found from

$$[E_{sys}]^T \begin{Bmatrix} \{U_g\} \\ - \\ \{0\} \end{Bmatrix} = [G_e]\{F_e\} \tag{6.43}$$

which we will write as $[E_{sys}]^T\{U_{sys}\} = [G_e]\{F_e\}$. Here is this solution in MATLAB:

The same parameters given previously are:

```
k = 4*10^6; p0 = 100; l = 6; P = 600;     % give parameters
g = 1/k;                                  % flexibility
```

Below is the extended global equilibrium matrix where we include the reactions.

and the corresponding force vector, assembled by hand:

(continued)

Example 6.2 (continued)

```
Ege = [ -1 0 0 0 0 0 -1 0; 1 -1 0 0 0 0 0 0; 0 1 -1 0 0 0 0 0;...
 0 0 1 -1 0 0 0 0; 0 0 0 1 -1 0 0 0; 0 0 0 0 1 -1 0 0;...
 0 0 0 0 0 1 0 -1];
Pe = [p0*l/2; p0*l; p0*l; p0*l/2; P; 0; 0];
```

Now, calculate the compatibility matrix from the extended equilibrium matrix:

```
S = null(Ege).';
```

and form the flexibility matrix with zero flexibilities for the reactions:

```
flex = g*ones(1,8);
Ge = diag(flex);
Ge(7,7) = 0; Ge(8,8) = 0;
```

Then form the system equations and solve for the forces:

```
Esys = [Ege; S*Ge];
Psys = [ Pe;0];
Fe = Esys\Psys

Fe =

 1250    % F1
 650     % F2
 50      % F3
 -250    % F4
 -850    % F5
 -850    % F6
 -1550   % R1
 -850    % R2
```

Finally, determine the displacements:

```
Usys = Esys.'\Ge*Fe

Usys =
 1.0e-03 *
```

(continued)

Example 6.2 (continued)

```
 0          % U1
0.3125      % U2
0.4750      % U3
0.4875      % U4
0.4250      % U5
0.2125      % U6
0           % U7
0           % extra zero
```

which gives the same solution as before.

6.3.1 Nonzero Specified Displacements in a Force-Based Method

We can also include non-zero displacements at supports, which is a case we discussed with a displacement-based approach. Again, we will use the problem of Fig. 6.3a as an example. Consider again the global equilibrium equation (Eq. (6.27):

$$\begin{bmatrix} -1 & 0 & 0 & 0 & 0 & 0 \\ 1 & -1 & 0 & 0 & 0 & 0 \\ 0 & 1 & -1 & 0 & 0 & 0 \\ 0 & 0 & 1 & -1 & 0 & 0 \\ 0 & 0 & 0 & 1 & -1 & 0 \\ 0 & 0 & 0 & 0 & 1 & -1 \\ 0 & 0 & 0 & 0 & 0 & 1 \end{bmatrix} \begin{Bmatrix} F_1 \\ F_2 \\ F_3 \\ F_4 \\ F_5 \\ F_6 \end{Bmatrix} = \begin{Bmatrix} R_1 + p_0 l/2 \\ p_0 l \\ p_0 l \\ p_0 l/2 \\ P \\ 0 \\ R_2 \end{Bmatrix} \tag{6.44}$$

and the corresponding deformation-displacement relations (Eq. (6.29))

$$\begin{Bmatrix} \Delta_1 \\ \Delta_2 \\ \Delta_3 \\ \Delta_4 \\ \Delta_5 \\ \Delta_6 \end{Bmatrix} = \begin{bmatrix} -1 & 1 & 0 & 0 & 0 & 0 & 0 \\ 0 & -1 & 1 & 0 & 0 & 0 & 0 \\ 0 & 0 & -1 & 1 & 0 & 0 & 0 \\ 0 & 0 & 0 & -1 & 1 & 0 & 0 \\ 0 & 0 & 0 & 0 & -1 & 1 & 0 \\ 0 & 0 & 0 & 0 & 0 & -1 & 1 \end{bmatrix} \begin{Bmatrix} U_1 \\ U_2 \\ U_3 \\ U_4 \\ U_5 \\ U_6 \\ U_7 \end{Bmatrix} \tag{6.45}$$

In the deformation-displacement relations, we remove the columns from the transpose of the equilibrium matrix that are associated with the specified global displacements at the supports, $\{U_s\}$, and place them in a matrix, $[E_R]^T$, giving

$$\{\Delta\} = [E_R]^T\{U_s\} + [E_f]^T\{U_f\} \tag{6.46}$$

where $[E_f]$ is the "free equilibrium matrix defined previously and $\{U_f\}$ are the "free" displacements. For our specific problem, for example, we have

$$\begin{Bmatrix} \Delta_1 \\ \Delta_2 \\ \Delta_3 \\ \Delta_4 \\ \Delta_5 \\ \Delta_6 \end{Bmatrix} = \begin{bmatrix} -1 & 0 \\ 0 & 0 \\ 0 & 0 \\ 0 & 0 \\ 0 & 0 \\ 0 & 1 \end{bmatrix} \begin{Bmatrix} U_1 \\ U_7 \end{Bmatrix} + \begin{bmatrix} 1 & 0 & 0 & 0 & 0 \\ -1 & 1 & 0 & 0 & 0 \\ 0 & -1 & 1 & 0 & 0 \\ 0 & 0 & -1 & 1 & 0 \\ 0 & 0 & 0 & -1 & 1 \\ 0 & 0 & 0 & 0 & -1 \end{bmatrix} \begin{Bmatrix} U_2 \\ U_3 \\ U_4 \\ U_5 \\ U_6 \end{Bmatrix} \tag{6.47}$$

Eliminating these columns in the deformation-displacement relations corresponds to removing the corresponding rows in the global equilibrium equations and the rows in the force vector that contain the unknown reaction forces, yielding $[E_f]$ $\{F\} = \{P_f\}$. Specifically

$$\begin{bmatrix} 1 & -1 & 0 & 0 & 0 & 0 \\ 0 & 1 & -1 & 0 & 0 & 0 \\ 0 & 0 & 1 & -1 & 0 & 0 \\ 0 & 0 & 0 & 1 & -1 & 0 \\ 0 & 0 & 0 & 0 & 1 & -1 \end{bmatrix} \begin{Bmatrix} F_1 \\ F_2 \\ F_3 \\ F_4 \\ F_5 \\ F_6 \end{Bmatrix} = \begin{Bmatrix} p_0 l \\ p_0 l \\ p_0 l/2 \\ P \\ 0 \end{Bmatrix} \tag{6.48}$$

Thus, at this stage, we have the equations

$$[E_f]\{F\} = \{P_f\}, \quad \{\Delta\} = [E_R]^T\{U_s\} + [E_f]^T\{U_f\}, \quad \{\Delta\} = [G]\{F\} \tag{6.49}$$

which is an extended form of the algebraic equations of elasticity. We again use the MATLAB null function to find a compatibility matrix $[S]$ that satisfies $[E_f][S]^T = 0$. Then Eq. (6.49) shows that we must have

$$[S]\{\Delta\} = [S][E_R]^T\{U_s\} \tag{6.50}$$

Writing the compatibility equations in terms of the internal forces through the flexibility matrix and appending these compatibility equations to the reduced equilibrium equations, we find

$$\begin{bmatrix} [E_f] \\ - \\ [S][G] \end{bmatrix} \{F\} = \left\{ \begin{array}{c} \{P_f\} \\ - \\ [S][E_R]^T\{U_s\} \end{array} \right\} \tag{6.51}$$

which simply reduces to the system we solved before as $[E_{sys}]\{F\} = \{P_{sys}\}$ but where the terms added to the force vector now are not zeros. After the internal forces are found, to solve for the displacements we start from the deformation-displacement relation written in terms of the forces:

$$[G]\{F\} = [E_R]^T\{U_s\} + [E_f]^T\{U_f\} \tag{6.52}$$

which we extend, as done before:

$$\begin{aligned} [G]\{F\} - [E_R]^T\{U_s\} &= \left[[E_f]^T \mid [G]^T[S]^T \right] \left\{ \begin{array}{c} \{U_f\} \\ - \\ \{0\} \end{array} \right\} \\ &= [E_{sys}]^T \left\{ \begin{array}{c} \{U_f\} \\ - \\ \{0\} \end{array} \right\} = [E_{sys}]^T\{U_{sys}\} \end{aligned} \tag{6.53}$$

which we can solve as the system of equations:

$$[E_{sys}]^T\{U_{sys}\} = [G]\{F\} - [E_R]^T\{U_s\} \tag{6.54}$$

for $\{U_{sys}\}$, which contains the free displacements and one or more zeros.

6.3.2 Stiffness Matrices

We saw in Chap. 2 that the algebraic equations of linear elasticity also can be reduced to a displacement-based formulation in terms of a pseudo-stiffness matrix. For the mth element, we have

$$\left[K_p^{(m)}\right] = \left[E^{(m)}\right]\left[G^{(m)}\right]^{-1}\left[E^{(m)}\right]^T \tag{6.55}$$

where $[G^{(m)}]^{-1} = k^{(m)}$, that is, it is just the stiffness of the element and recall the element equilibrium matrix is

$$\left[E^{(m)}\right] = \begin{bmatrix} -1 \\ 1 \end{bmatrix} \tag{6.56}$$

so that carrying out the matrix multiplication gives

$$\left[K_p^{(m)}\right] = \begin{bmatrix} k^{(m)} & -k^{(m)} \\ -k^{(m)} & k^{(m)} \end{bmatrix} \tag{6.57}$$

which is the ordinary stiffness matrix for a bar element. Thus, solving axial extension problems with the pseudo stiffness matrix is the same as with the stiffness matrix in the traditional displacement-based finite element method, as we have noted previously. This is also true for torsion problems and beam bending problems. For the torsion problem, we again have Eq. (6.57) with $k^{(m)} = G^{(m)}J^{(m)}/l_{(m)}$, where $G^{(m)}$ is the shear modulus, $J^{(m)}$ is the polar area moment of the cross-section of the mth bar element, and $l_{(m)}$ is the element length. The element equilibrium matrix is the same as for the axial extension of a bar. For beam bending, it can be shown that [2].

$$\left[K^{(m)}\right] = \left[K_p^{(m)}\right] = \frac{Y^{(m)}I^{(m)}}{l_{(m)}^3} \begin{bmatrix} 12 & 6l_{(m)} & -12 & 6l_{(m)} \\ 6l_{(m)} & 4l_{(m)}^2 & -6l_{(m)} & 2l_{(m)}^2 \\ -12 & -6l_{(m)} & 12 & -6l_{(m)} \\ 6l_{(m)} & 2l_{(m)}^2 & -6l_{(m)} & 4l_{(m)}^2 \end{bmatrix} \tag{6.58}$$

where $Y^{(m)}$ is Young's modulus and $I^{(m)}$ is an area moment of the cross-sectional area of the beam. One can also show that an element equilibrium matrix for beam bending is [2].

$$\left[E^{(m)}\right] = \begin{bmatrix} -\dfrac{1}{l_{(m)}} & \dfrac{1}{l_{(m)}} \\ -1 & 0 \\ \dfrac{1}{l_{(m)}} & -\dfrac{1}{l_{(m)}} \\ 0 & 1 \end{bmatrix} \tag{6.59}$$

so that these stiffness and equilibrium matrices form the basis of displacement-based and force-based finite element methods for beams. In this book, we will not consider those cases.

6.4 Finite Element Scripts

6.4.1 Assembly Process in Finite Elements

Finite element solutions were obtained in the previous sections by forming the equilibrium or stiffness matrices without programming, but this was possible only because the elements and nodes were numbered sequentially, making the assembly process straightforward. In general, such sequential numbering is not possible, so a general assembly process is needed, using programming tools. To illustrate this process, we will again use members under axial loads, but where now there are multiple elastic bars attached to each other through rigid connectors that move in smooth guides that keep the connectors horizontal so that only force equilibrium in the x-direction need be considered (see Fig. 6.8 where the rigid connectors are shown as black members, and the guides are shown as dotted lines). The entire assembly will be fixed at both ends so the problem in Fig. 6.8 is statically indeterminate. Element numbers for the bars are given in parentheses and the node numbers are given without parentheses. All the applied forces will be assumed to be applied at the connectors, which are also the nodes. As can be seen in Fig. 6.8 the numbering pattern for both the elements and the nodes can be chosen freely without any predetermined rule. To define the specific numbering pattern used, we will employ a *connectivity matrix* that defines the appropriate node numbers for each element. For example, for the bars in Fig. 6.8, we can form the connectivity matrix shown in Fig. 6.9, where each row number represents the element number, the number in the first column represents the starting node number for the element, and the number in the second column represents the ending node number for the element. The connectivity matrix gives us all the information we need to form a global equilibrium matrix or global stiffness matrix. In a displacement-based finite element method, the connectivity matrix is also needed to determine the forces in the bar elements, which are related to the displacements at the end nodes.

Consider, for example, forming a global equilibrium matrix for the bar assembly shown in Fig. 6.8. Figure 6.10b shows the mth bar element for such assembly and the equilibrium matrix for that mth element. The mth element equilibrium matrix is a 2×1 matrix $E_e(m,n)$ $(m = 1,2)$, $n = 1$, where $E_e(1,1) = -1$, $E_e(2,1) = 1$. When placed in a global equilibrium matrix this bar element contribution will be placed as shown in Fig. 6.11, which also shows the force in the bar element in the vector of unknown internal forces and the forces at the end nodes.

Now, consider the MATLAB code for the assembly process. Let us use the bar assembly of Fig. 6.8 as an example. Let Ne = 6 be the number of bars and Nn = 5 the number of nodes, and let's define a connectivity matrix as a Ne × 2 matrix named connect containing the appropriate node numbers for those bars. The element equilibrium matrix in MATLAB is defined as a matrix Ee and the global equilibrium matrix Eg will be a Nn × Ne matrix, which we will initially populate with zeros:

```
Ne = 6;
Nn = 5;
connect = [1 3; 5 4; 4 2; 3 4; 3 5; 3 5];

Ee = [-1;1]
Eg = zeros(Nn, Ne)
```

We then can assemble the global equilibrium matrix through the following program steps:

```
for nn = 1:Ne
nstart = connect(nn,1);
nend = connect(nn,2);
Eg(nstart, nn) = Eg(nstart,nn) + Ee(1,1);
Eg(nend,nn) = Eg(nend,nn) + Ee(2,1);
end
```

Consider now forming a global stiffness matrix for the bars shown in Fig. 6.8. Figure 6.10c shows the mth bar element and the stiffness matrix for that mth element. The mth stiffness matrix is a 2 × 2 matrix $K_e(m,n)$, ($m = 1,2$), ($n = 1,2$) where $K_e(1,1) = K_e(2,2) = k^{(m)}$, $K_e(1,2) = K_e(2,1) = -k^{(m)}$ for the mth element, where $k^{(m)}$ is the stiffness of the element. When placed in a global equilibrium matrix this bar element contribution will be placed as shown in Fig. 6.12, which also shows the end displacements in the bar in the vector of unknown displacements and the forces at the end nodes.

For specificity, assume that all the stiffnesses are the same and equal to $k^{(m)} = 120$ kN/m.

Using the bars of Fig. 6.8 again, Ne = 6 is the number of bars, Nn = 5 is the number of nodes, and the connectivity matrix is the same Ne × 2 matrix previously named connect containing the appropriate node numbers for those bars. The global stiffness matrix will be a Nn × Nn matrix, which we will initially populate with zeros. The MATLAB code for the assembly process is then:

```
kconstant =120*ones(1,Ne)   % all element stiffnesses are the same
Ne = 6;
Nn = 5;
connect = [1 3; 5 4; 4 2; 3 4; 3 5; 3 5];
Kg = zeros (Nn, Nn)
```

The assembly process is then

```
for nn = 1:Ne
ke = kconstant(nn);            % element spring constant for element nn
Ke = [ ke -ke; -ke ke];
nstart = connect(nn,1);        % element beginning node
nend = connect(nn,2);          % element finishing node
Kg(nstart, nstart) = Kg(nstart,nstart) + Ke(1,1);
Kg(nstart, nend) = Kg(nstart,nend) + Ke(1,2);
Kg(nend, nstart) = Kg(nend,nstart) + Ke(2,1);
Kg(nend,nend) = Kg(nend,nend) + Ke(2,2);
end
```

There are obviously other programming aspects needed in finite element programs, but assembly of the elements of the global equilibrium or stiffness matrix is a key element that we wanted to highlight. For the other details, we will use MATLAB scripts in the following sections.

6.4.2 Force-Based Finite Element Script

The MATLAB script FEM_force_bars solves the system of bars in Fig. 6.8 for the forces and displacements with a force-based approach. In the script, all the bar stiffnesses are given by $k = 120$ kN/m so all the element flexibilities will also be the same. Since the bars are not all the same lengths, this means that the *EA* in the stiffness expressions $k = EA/l$ must be such that they compensate for the length differences. This choice was made only to keep the user inputs as simple as possible in the example being discussed and can be easily changed. There is only one applied force $P_5 = 20$ kN at node 5. The MATLAB script content will be described line-by-line interspersed with MATLAB and non-MATLAB comments and explanations. Here is the script and commentary:

The first section of the script provides all the data inputs needed to solve for the forces and displacements.

```
% script FEM_force_bars

% USER PROVIDED INPUTS
% ______________________________________________

Nn = 5;   % no. of nodes
Ne = 6;   % no. of elements (also no. of internal forces for this problem)
kconstant =120*ones(1,Ne)        % bar stiffnesses in kN/m
connect = [1 3; 5 4; 4 2; 3 4; 3 5; 3 5];   % connectivity matrix
```

```
% displacement and force boundary conditions will both be given
% as [node no. value]

bcd = [ 2 0 ; 1 0]; % displacement boundary conditions - must go from high
% to low node no.
Ndbc = size(bcd, 1); % determine no. of displacement bcs, Ndbc

% make sure the displacement bcs node numbers are in descending order,
% which is necessary when removing equilibrium equations later

for nn = 1:Ndbc-1
if bcd(nn+1,1) > bcd(nn,1)
error('displacement bc node numbers must be in descending order')
end
end

bcf = [5 20]; % force boundary conditions, kN
Nfbc = size(bcf,1); % determine no. of force boundary conditions, Nfbc

% END OF USER PROVIDED INPUTS
% ______________________________________________________________
```

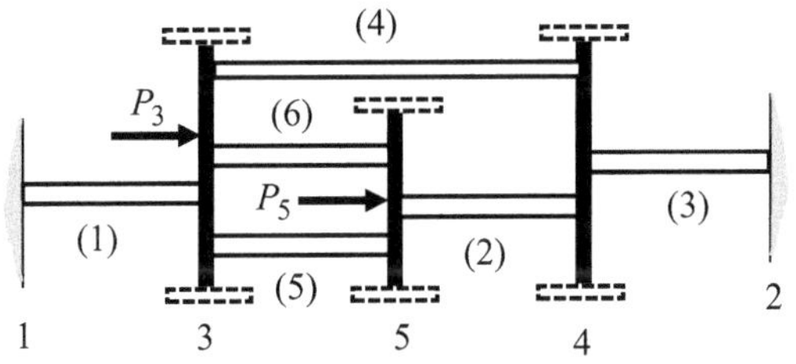

Fig. 6.8 An assembly of elastic bars connected by rigid members that ride in smooth guides (dashed lines) to keep them horizontal. The entire assembly is fixed at both ends. The bar elements are the numbers given in parentheses and the bar nodes are the numbers given without parentheses

Fig. 6.9 A connectivity matrix for the bar assembly in Fig. 6.8

element	node numbers	
(1)	1	3
(2)	5	4
(3)	4	2
(4)	3	4
(5)	3	5
(6)	3	5

connectivity matrix

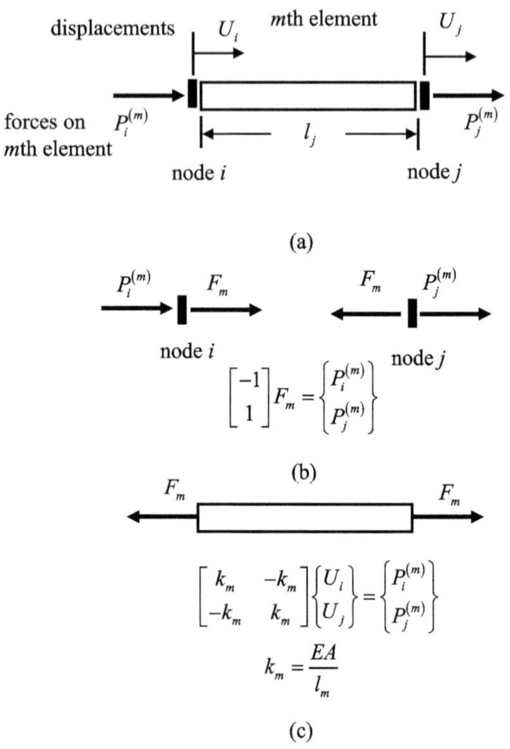

Fig. 6.10 (**a**) The *m*th bar element. (**b**) The equilibrium matrix for the *m*th element. (**c**) The stiffness matrix for the *m*th element

Next, we define the flexibility matrix and the element equilibrium matrix and assemble the global equilibrium matrix.

```
G = diag(1./kconstant); % flexibility matrix. Note the use of ./
Ee = [-1;1]; % element equilibrium matrix
Eg = zeros(Nn, Ne); % initialize the global equilibrium matrix to all
zeros

% assemble the global equilibrium matrix
for nn = 1:Ne
nstart = connect(nn,1);
nend = connect(nn,2);
Eg(nstart, nn) = Eg(nstart,nn) + Ee(1,1);
Eg(nend,nn) = Eg(nend,nn) + Ee(2,1);
end
```

Now, apply the force and displacement boundary conditions. The displacement boundary conditions will be satisfied by generating a reduced equilibrium matrix and force vector where the rows are eliminated from the global equilibrium matrix and global force vector where the displacements are specified (and where the unknown

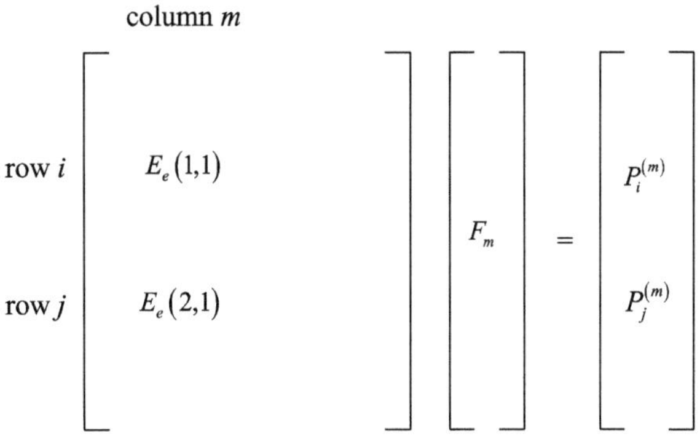

Fig. 6.11 Placement of element equilibrium matrix components in a global equilibrium matrix

reaction forces are present). The eliminated rows in the global equilibrium matrix will be saved in the Er matrix, which is needed to include any nonzero specified displacements, as described previously in this chapter, and will be used to generate the reaction forces at the nodes where the displacements are specified. Thus, in this script, we will allow for the specified displacements to be non-zero values, but our specific example will be with only zero displacements (see user inputs).

```
P = zeros(Nn,1); % initialize global force vector initially with zeros

% apply the force boundary conditions
for ii = 1: Nfbc
P(bcf(ii, 1)) = bcf(ii,2);
end

% apply the displacement boundary conditions and obtain
% the reduced equilibrium matrix, E, and reduced loads
% vector, P, and matrix of removed equations, Er, and
% specified displacements vector, Us.

Er = zeros(Ndbc, Ne); % initialize the matrix of removed equations
Us = zeros(Ndbc,1); % initialize a vector holding known displacements
E = Eg; % E will be the reduced equilibrium matrix
% generate the reduced equilibrium matrix E, the reduced load vector P,
%and the matrix of removed equilibrium equations, Er, and generate a
% vector Us with the given displacements

for jj = 1:Ndbc
Er(jj,:) = Eg(bcd(jj,1), :),
Us(jj) = bcd(jj,2);
E (bcd(jj,1), :), = [ ];
P(bcd(jj,1)) = [ ];

end
```

$$
\begin{array}{c} \\ \text{row } i \\ \\ \text{row } j \\ \\ \end{array}
\begin{array}{c}
\begin{array}{cc} \text{column } i & \text{column } j \end{array} \\
\begin{bmatrix} & & \\ & K_e(1,1) & K_e(1,2) \\ & & \\ & K_e(2,1) & K_e(2,2) \\ & & \end{bmatrix}
\end{array}
\begin{bmatrix} \\ U_i \\ \\ U_j \\ \\ \end{bmatrix}
=
\begin{bmatrix} \\ P_i^{(m)} \\ \\ P_j^{(m)} \\ \\ \end{bmatrix}
$$

Fig. 6.12 Placement of element stiffness matrix components in a global stiffness matrix

E is now the reduced equilibrium matrix and P is the reduced load vector. Next, we need to generate the compatibility matrix, form the compatibility equations in terms of the internal forces, and append those compatibility equations to the reduced equilibrium equations, which we can then solve for the internal forces:

```
% generate the compatibility matrix and compatibility eqs.
S = null(E).';
C = S*G; % multiply compatibility matrix by flexibility matrix
Ncomp =size(C,1); % determine the number of compatibility
% equations, Ncomp.
% Set up the combined system of modified equilibrium and compatibility
% equations and solve for the internal forces, F
Esys = [Eg; C];
Psys = [P; S*Er'*Us];
F = Esys\Psys; % solve system for internal forces
```

Next, solve for the free displacements (and some zeros) and combine those free displacements with the specified displacements to generate the global displacement vector:

```
Usys = Esys'\(G*F- Er'*Us); %solve for the system displacements, Usys
Uf = Usys(1:Ne - Ncomp); % extract the free displacements and put the
% free and specified displacements into global displacement vector
Ug. The
% number of free displacements is Ne - Ncomp since Ne is also the
% number of internal elongations (and internal forces) and the number of
% compatibility equations is the number of relations between
elongations
% when those elongations are expressed in terms of the free displacements
```

```
Ug = 1i*ones(Nn,1); % initialize Ug with all imaginary values
% replace imaginary values with the known displacements
% at specified nodes
for nn = 1:Ndbc
Ug(bcd(nn,1)) = bcd(nn,2);
end
% replace imaginary values remaining with free displacements, Uf
kk = 1; % initialize counter
for jj = 1:Nn
if imag(Ug(jj)) == 1
Ug(jj) = Uf(kk);
kk = kk+1;
end
end
```

Also compute the reactions at the nodes where the displacements are satisfied. Note that we have only simulated a system of 1-D elements where the forces are applied at the nodes. For cases where there are equivalent nodal forces acting at the nodes where the reactions act, those forces will also have to be included.

```
% determine the reactions
R = Er*F;
```

Finally, we display all the results:

```
% display internal forces, F , for all elements
disp("The internal bar forces are")
disp(F)
% display global displacements, Ug, at all nodes
disp("The global displacements are")
disp(Ug)

% display the reaction forces and their nodes
disp ("At the supports where the node numbers are")
node_numbers = bcd(1:Ndbc,1);
disp(node_numbers)
disp("the reactions are")
disp(R)
```

If we run this script, the outputs are.

```
FEM_force_bars
The internal bar forces are
 10.7692
```

```
 -7.6923
 -9.2308
 -1.5385
  6.1538
  6.1538
The global displacements are
      0
      0
  0.0897
  0.0769
  0.1410
At the supports where the node numbers are
   2
   1
the reactions are
  -9.2308
 -10.7692
```

These forces and displacements are shown in Fig. 6.13. The forces are measured in kN and the displacements measured in m.

It is useful to examine the compatibility equation in this problem. The reduced equilibrium matrix found in the script is:

```
E
E =

   1   0   0   -1 -  1   -1
   0   1  -1   1   0   0
   0  -1   0   0   1   1
```

If we compute the compatibility matrix with the null function and the 'r' option, we find.

```
Sc = null(E,'r').'
Sc =

   1   0   1   1   0   0
   1   1   1   0   1   0
   1   1   1   0   0   1
```

The compatibility equations for the deformations (elongations) are $[S]\{\Delta\} = 0$ (note: we did not use the 'r' option in the script so the compatibility equations there will look different). Another way to generate these compatibility equations is to observe that the elongations must not produce a change of length for any sequence of bars between the two fixed supports. There are three independent ways in which we can consider the sum of the elongations equal to zero between the two fixed supports for this system (see Fig. 6.8):

$$\Delta_1 + \Delta_3 + \Delta_4 = 0$$
$$\Delta_1 + \Delta_2 + \Delta_3 + \Delta_5 = 0$$
$$\Delta_1 + \Delta_2 + \Delta_3 + \Delta_6 = 0$$

or, equivalently,

$$\begin{bmatrix} 1 & 0 & 1 & 1 & 0 & 0 \\ 1 & 1 & 1 & 0 & 1 & 0 \\ 1 & 1 & 1 & 0 & 0 & 1 \end{bmatrix} \begin{Bmatrix} \Delta_1 \\ \Delta_2 \\ \Delta_3 \\ \Delta_4 \\ \Delta_5 \\ \Delta_6 \end{Bmatrix} = \begin{Bmatrix} 0 \\ 0 \\ 0 \end{Bmatrix}$$

so that the results from the null function and from our physical argument are the same. This result could be used, as well as the normalized compatibility matrix found in the script. We have illustrated the force-based finite element method with a very simple 1-D system. For more details on general 3-D implementations, see [3, 4], where the method is called the integrated force method.

6.4.3 Displacement-Based Finite Element Script

The MATLAB script FEM_stiffness_bars solves the system of bars in Fig. 6.8 for the forces and displacements with a displacement-based approach. In the script, all inputs are the same as in the previous force-based case. The bar stiffnesses are given by $k = 120$ kN/m so all the element flexibilities will also be the same. There is only one applied force $P_5 = 20$ kN at node 5. The MATLAB script content will be discussed line-by-line interspersed with MATLAB and non-MATLAB comments and explanations. Here is the script and commentary:

The user inputs are very similar to the force-based case. However, in this solution, we will not be eliminating equations from the global equations, so the ordering of the displacement boundary conditions is immaterial.

```
% script FEM_stiffness_bars

% USER PROVIDED INPUTS
% ______________________________________________________________
Nn = 5; % number of nodes (also number of global displacements)
```

```
Ne = 6; % number of elements
% spring constants for the elements (must be of length Ne)
kconstant = 120*ones(1, Ne);
% connectivity matrix, where each row has the values
% [start_node_number end_node_number] . It must be a matrix of (Ne x 2)
dimensions
connect = [1 3; 3 4; 3 5; 3 5; 5 4; 4 2];
% matrix of displacement boundary conditions, bcd, and matrix of
% force boundary conditions, bcf, where each row contains
% [global_unknown _number value ]
bcd = [1 0; 2 0];
bcf = [5 20];
Nbcd = size(bcd,1) % number of displacement boundary conditions
Nbcf = size(bcf,1); % number of force boundary conditions
% END OF USER PROVIDED INPUTS
% ________________________________________________________________
```

The next steps are those involved in assembling the global stiffness matrix and modifying it. We will not be eliminating equations that contain the reaction forces (at the nodes where the displacements are specified). Instead, those equations will be modified so we will need to identify those modified equations before the modification so we can determine the reaction forces later. Basically, the modifications involve turning those equations with reaction forces into equations that specify the given displacements, as discussed earlier.

```
% initialize the element stiffness matrix, Ke, and the global stiffness
% matrices, Kg, with % zeros. Also, initialize a matrix Kr to hold the
% rows of the global stiffness matrix that will % generate the reaction
forces

Ke = zeros(2,2);
Kg = zeros(Nn , Nn);
Kr = zeros(nbcd,Nn);
% Generate the element stiffness matrices and assemble into
% the global stiffness matrix
% Use the connectivity matrix to place the element stiffness
% values in their correct positions

for nn = 1:Ne
ke = kconstant(nn);          % element spring constant
Ke = [ ke -ke; -ke ke];
beg = connect(nn,1);         % element beginning node
fin= connect(nn,2);          % element finishing node
Kg(beg, beg) = Kg(beg,beg) + Ke(1,1);
Kg(beg, fin) = Kg(beg,fin) + Ke(1,2);
```

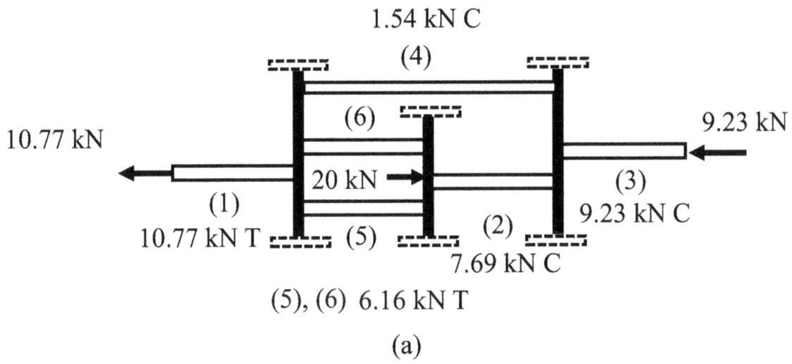

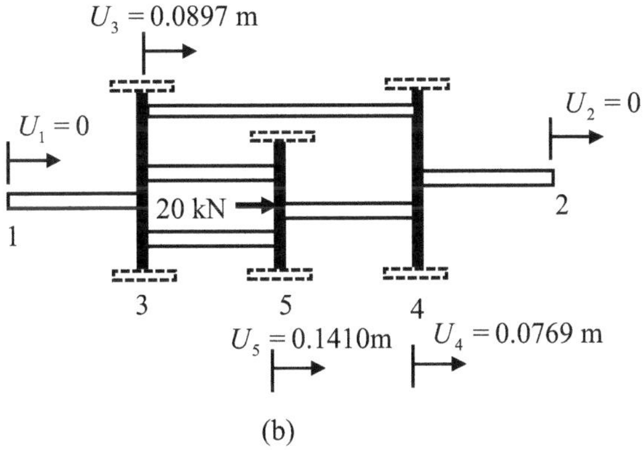

Fig. 6.13 (**a**) The forces acting on the system of bars shown in Fig. 6.8. (**b**) The displacements of the bar system

```
Kg(fin, beg) = Kg(fin,beg) + Ke(2,1);
Kg(fin,fin) = Kg(fin,fin) + Ke(2,2);
end

% initialize the force vector

P = zeros(Nn, 1);

% apply the displacement boundary conditions and save the rows
% needed to evaluate the reactions with the matrix Kr
 [maxK , i] = max(Kg(:)); % find largest stiffness
for ii = 1:Nbcd
node = bcd(ii,1);
Kr(ii,:) = Kg(node,:);
Kg(node,node) = maxK*1.0E9;
P(node) = bcd(ii,2)*maxK*1.0E9;
end
```

```
% apply the force boundary conditions
for ii = 1: Nbcf
node = bcf(ii,1);
P(node) = bcf(ii,2);
end
```

We can then solve for the displacements:

```
% solve the system of equations
u = Kg\P;

% evaluate internal forces in the bars
F = zeros(Ne,1); % initialize the force vector
```

Calculating the internal bar forces is done by again using the connectivity matrix to find the finish and start points of an element and then the force is just the stiffness of the element times the difference of displacements at those points:

```
for jj= 1:Ne
k = kconstant(jj);
start = connect(jj,1);
finish = connect(jj, 2);
F(jj) = k*(u(finish) - u(start));
end
% display all the results
disp ("The global displacements are:")
disp(u)

disp (" The internal bar forces are:")
disp(F)
```

The Kr matrix times the global displacements gives the reaction forces at the nodes where the displacements were specified. This assumes that there are no equivalent forces acting at these nodes, as is the case here since we assumed all the applied forces are nodal forces.

```
disp ("At the supports where the node numbers are:")
node_numbers = bcd(1:Nbcd,1);
disp(node_numbers)
disp("the reactions are")
R = Kr*u;
disp(R)
```

For the system of bars being considered, running the script gives the following results, where the forces are measured in kN and the displacements measured in m:

```
FEM_stiffness_bars
The global displacements are:
   0.0000
   0.0000
   0.0897
   0.0769
   0.1410

 The internal bar forces are:
 10.7692
 -7.6923
 -9.2308
 -1.5385
 6.1538
 6.1538

At the supports where the node numbers are:
   1
   2

the reactions are
  -10.7692
  -9.2308
```

These results agree with the force-based solution. Again, we have illustrated the displacement-based finite element method with a simple 1-D example. The next section will briefly examine more general cases. For details of a more general implementation, see, for example, [5].

6.5 Finite Elements in More General Cases

This chapter examined finite elements for only 1-D problems so that we could better illustrate some of the main features of the method. When elements are present in higher dimensional problems the element equilibrium and stiffness matrices become more complex, but the basic steps remain the same. Consider, for example, truss problems. In a force-based finite element method, we need to obtain the element equilibrium matrix for the *n*th element that is at a general angle θ_n with to respect to the *x*-axis, as shown in Fig. 6.14a. From the free body diagrams at the end nodes (Fig. 6.14b), we have

$$[E]\{F\} = \{P\} \quad \rightarrow \quad \begin{bmatrix} -\cos(\theta_n) \\ -\sin(\theta_n) \\ \cos(\theta_n) \\ \sin(\theta_n) \end{bmatrix} \{F_n\} = \begin{Bmatrix} P_{xi}^{(n)} \\ P_{yi}^{(n)} \\ P_{xj}^{(n)} \\ P_{yj}^{(n)} \end{Bmatrix} \tag{6.60}$$

for the bar force F_n in terms of the Cartesian components of the forces acting at the end nodes (i, j). For a truss, these element equilibrium matrices can be assembled into a global equilibrium matrix, as done before, and the force-based solution obtained. We can also define a pseudo stiffness matrix for a truss element. If we let $c = \cos(\theta_n)$, $s = \sin(\theta_n)$, and let k be the member stiffness, in MATLAB, we have.

```
syms c s k
E = [-c; -s; c; s];
G = 1/k;
Kp = E*inv(G)*E.'

Kp =

[ c^2*k, c*k*s, -c^2*k, -c*k*s]
[ c*k*s, k*s^2, -c*k*s, -k*s^2]
[-c^2*k, -c*k*s, c^2*k, c*k*s]
[-c*k*s, -k*s^2, c*k*s, k*s^2]
```

so the element pseudo stiffness matrix (which is the same as the ordinary stiffness matrix) is

$$[K_p] = k \begin{bmatrix} \cos^2\theta_n & \cos\theta_n \sin\theta_n & -\cos^2\theta_n & -\cos\theta_n \sin\theta_n \\ \cos\theta_n \sin\theta_n & \sin^2\theta_n & -\cos\theta_n \sin\theta_n & -\sin^2\theta_n \\ -\cos^2\theta_n & -\cos\theta_n \sin\theta_n & \cos^2\theta_n & \cos\theta_n \sin\theta_n \\ -\cos\theta_n \sin\theta_n & -\sin^2\theta_n & \cos\theta_n \sin\theta_n & \sin^2\theta_n \end{bmatrix} \tag{6.61}$$

and the equilibrium equations for the nth element in terms of the Cartesian nodal displacement components (see Fig. 6.14c) are

$$[K_p] \begin{Bmatrix} U_{xi}^{(n)} \\ U_{yi}^{(n)} \\ U_{xj}^{(n)} \\ U_{yj}^{(n)} \end{Bmatrix} = \begin{Bmatrix} P_{xi}^{(n)} \\ P_{yi}^{(n)} \\ P_{xj}^{(n)} \\ P_{yj}^{(n)} \end{Bmatrix} \tag{6.62}$$

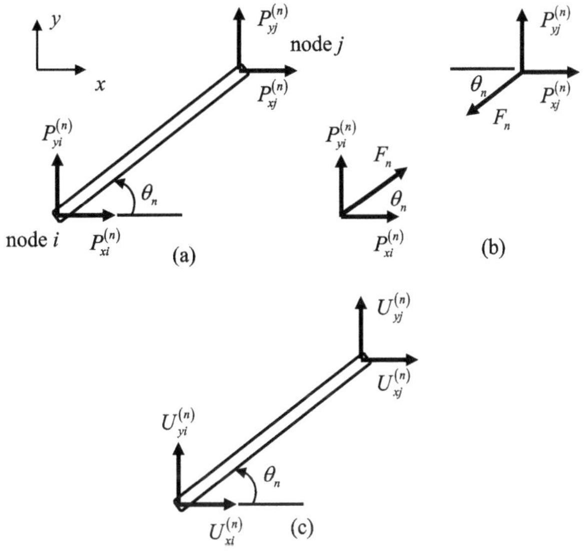

Fig. 6.14 (**a**) A truss element and the forces acting on it. (**b**) Free body diagrams at the pins (nodes). (**c**) The pin (nodal) displacements

which then can be assembled into a global stiffness matrix and solved for the displacements. These are more complex equilibrium and stiffness matrices because the geometry information is embedded in those matrices. In contrast, when solving problems on a node-by-node basis, the geometry information is explicitly given at the nodes. For beam and frame problems, the element matrices are even more complex but once they are defined, they can be assembled via programming steps like those shown for 1-D problems. We will not show those details here.

References

1. Kurrer, K.-E. (2018). *The history of the theory of structures – Searching for equilibrium* (2nd ed.). Ernst & Sohn.
2. Schmerr, L. W. (2021). *Advanced mechanics of solids*. Cambridge University Press.
3. Kaljević, I., Patnaik, S. N., & Hopkins, D. A. (1996). *Development of finite elements for two-dimensional structural analysis using the integrated force method* (NASA Technical Memorandum 4655). NASA, Lewis Research Center.
4. Kaljević, I., Patnaik, S. N., & Hopkins, D. A. (1996). *Element library for three-dimensional stress analysis by the integrated force method* (NASA Technical Memorandum 4686). NASA, Lewis Research Center.
5. Li, G. (2020). *Introduction to the finite element method and implementation with MATLAB*. Cambridge University Press.

Appendix A: MATLAB® Functions

Force-based methods can be implemented using only the built-in functions of MATLAB but there are two areas where additional functions can be extremely useful. Singularity functions, for example, make the evaluation and plotting of internal forces and moments much easier, and a special function for evaluating equivalent nodal forces and moments can avoid having to write out rather complex expressions. This appendix gives the listings of those supplemental MATLAB functions. The MATLAB m-files of these supplemental functions and the finite element scripts described in Chap. 6 can be obtained by writing an e-mail with the subject title Algebraic Equation Codes to the author at lschmerr@iastate.edu.

Singularity Functions

Here are more detailed listings of the singularity functions used for bending problems. The functions can also be used for extension and torsional problems.

step singularity function

```
function y = step_sf(x,a)
% y = step_sf(x, a) generates a singularity function that is zero for x < a
% and is one for x > a. For a concentrated load P the shear
% force is V = - P*step_sf(x,a) while for a concentrated couple C
% we have the bending moment M = C*step_sf(x,a).

if x(1) == a      % if the step is located at the first point then let the
  y = (x>a);      % value at the first point be zero, otherwise
else              % let that value be unity. This allows MATLAB to plot
  y = (x>= a);  % the jump in the function properly.
end
```


L. W. Schmerr Jr., *Algebraic Equations of Linear Elasticity*,
https://doi.org/10.1007/978-3-031-66174-7

linear singularity function

```
function y = lin_sf(x,a)
% y = lin_sf(x, a) generates a singularity function that is zero for x < a
% and varies like (x - a) for x > a. For a concentrated load P the bending
% moment is M = P*lin_sf(x,a) while for a uniform load pz that starts at
% x = a we have the shear force V = - pz*lin_sf(x,a).

y = (x -a).*(x > a);
```

quadratic singularity function

```
function y = quad_sf(x,a)
% y = quad_sf(x, a) generates a singularity function that is zero for x < a
% and varies like (x - a).^2 for x > a. For a uniform distributed load
% of intensity pz force/unit length that starts at x = a, the bending
moment
% is M = (pz/2)*quad_sf(x,a) while for a linearly increasing distributed
load
% whose slope is s force/unit length^2 the shear force is
% V = - (s/2)*quad_sf(x,a).

y = ((x-a).^2).*(x >a);
```

cubic singularity function

```
function y = cubic_sf(x,a)
%y = cubic_sf(x, a) generates a singularity function that is zero for x <
a
% and varies like (x - a).^3 for x > a. for a linearly increasing
% distributed load that starts at x = a and whose slope is
% s force/unit length^2 the bending moment is M = (s/6)*cubic_sf(x,a).

y = ((x-a).^3).*(x > a);
```

Equivalent Forces/Moments for Beams

This MATLAB function generates the equivalent nodal forces/moments for the loadings described in Fig. 5.12.

equiv_loads function

```
function [Fa, Ma, Fb, Mb] = equiv_loads(p, a, L,type)
% [Fa, Ma, Fb, Mb] = equiv_loads(p, a , L , type) generates the
% the equivalent forces and moments for distributed and concentrated
% forces. For distributed loads p is the force/unit length, a is the
% extent of the load, and L is the total element length. For a concentrated
% force p is the force and a is where the load acts in an element of
% length L. The argument type is a string given by:
% 'uniform', 'linear', or 'conc'

loading = type;
switch loading

  case{'uniform'}
  Fa = p*a*(2*L^3 -2*L*a^2 + a^3)/(2*L^3);
  Fb = p*a^3*(2*L-a)/(2*L^3);
  Ma = p*a^2*(6*L^2 - 8*L*a + 3*a^2)/(12*L^2);
  Mb = p*a^3*(4*L - 3*a)/(12*L^2);

  case {'linear'}
  Fa = p*a*(10*L^3 - 15*L*a^2 + 8*a^3)/(20*L^3);
  Fb = p*a^3*(15*L - 8*a)/(20*L^3);
  Ma = p*a^2*(10*L^2 - 15*L*a + 6*a^2)/(30*L^2);
  Mb = p*a^3*(5*L - 4*a)/(20*L^2);

  case {'conc'}
  Fa = p*(1 - 3*a^2/L^2 + 2*a^3/L^3);
  Fb = p*(a^2/L^2)*(3 - 2*a/L);
  Ma = p *(a - 2*a^2/L + a^3/L^2);
  Mb = p*(a^2/L - a^3/L^2);

  otherwise
    disp('Error. type must be ''uniform'', ''linear'', or ''conc''')
end
```

Index

L. W. Schmerr Jr., *Algebraic Equations of Linear Elasticity*, https://doi.org/10.1007/978-3-031-66174-7

MIX
Papier aus verantwortungsvollen Quellen
Paper from responsible sources
FSC® C105338
FSC
www.fsc.org

If you have any concerns about our products,
you can contact us on
ProductSafety@springernature.com

In case Publisher is established outside the EU,
the EU authorized representative is:
Springer Nature Customer Service Center GmbH
Europaplatz 3, 69115 Heidelberg, Germany

Printed by Libri Plureos GmbH
in Hamburg, Germany